Studies in Computational Intelligence

Volume 715

Series editor

Janusz Kacprzyk, Polish Academy of Sciences, Warsaw, Poland
e-mail: kacprzyk@ibspan.waw.pl

About this Series

The series "Studies in Computational Intelligence" (SCI) publishes new developments and advances in the various areas of computational intelligence—quickly and with a high quality. The intent is to cover the theory, applications, and design methods of computational intelligence, as embedded in the fields of engineering, computer science, physics and life sciences, as well as the methodologies behind them. The series contains monographs, lecture notes and edited volumes in computational intelligence spanning the areas of neural networks, connectionist systems, genetic algorithms, evolutionary computation, artificial intelligence, cellular automata, self-organizing systems, soft computing, fuzzy systems, and hybrid intelligent systems. Of particular value to both the contributors and the readership are the short publication timeframe and the worldwide distribution, which enable both wide and rapid dissemination of research output.

More information about this series at http://www.springer.com/series/7092

Ronald R. Yager · Jordán Pascual Espada
Editors

New Advances in the Internet of Things

Springer

Editors
Ronald R. Yager
Machine Intelligence Institute
Iona College
New Rochelle, NY
USA

Jordán Pascual Espada
Department of Computer Science
Oviedo University
Oviedo
Spain

ISSN 1860-949X ISSN 1860-9503 (electronic)
Studies in Computational Intelligence
ISBN 978-3-319-86343-6 ISBN 978-3-319-58190-3 (eBook)
DOI 10.1007/978-3-319-58190-3

Softcover reprint of the hardcover 1st edition 2017

Printed on acid-free paper

This Springer imprint is published by Springer Nature
The registered company is Springer International Publishing AG
The registered company address is: Gewerbestrasse 11, 6330 Cham, Switzerland

Preface

Internet of Things (IoT) promotes the creation of systems based on interconnected objects. These objects combine a physical and an electronic part, and they could be embedded devices, connected sensors, and other kind of electronic machines or devices. Most of IoT objects include sensors and actuators for interacting with the physical world.

The communication skill is one the main features of these objects. IoT objects use different communication protocols such as Internet or Bluetooth for exchanging information among them and other systems in an autonomous way without human interaction. Collaboration among objects is highly useful to improve tasks and processes in a wide range of different areas such as smart cities, logistic, transportation, smart homes, health systems, smart manufacturing, wearables, logistics, and agriculture. These IoT systems already provide great benefits to our society, but there are still a lot of challenges, improvement options, and new application fields.

This book was edited by Ronald R. Yager and Jordán Pascual. It selects ten chapters about recent and significant advances in the field of Internet of Things. Chapters are related to key fields: (1) RFID sensor networks and industrial Internet of Things; (2) communication efficiency in IoT networks; (3) publish/subscribe wireless sensor networks; (4) security and data quality; (5) smart cities, collective intelligence, and the Web of Things; (6) real-time protocols, wireless communication, and congestion control; (7) smart connectivity and user centric IoT applications; (8) storage system for data analytics in IoT; (9) IoT platforms, network protocols, and Quality of Service; and (10) mobile nodes in wireless networks. This selection presents ten relevant and innovative trends which will be able to generate great benefits over the current IoT development.

New Rochelle, USA | Ronald R. Yager
Oviedo, Spain | Jordán Pascual Espada

Contents

RFID-Based Multi-level Sensing Network for Industrial Internet of Things 1
S. Amendola, C. Occhiuzzi, S. Manzari and G. Marrocco

Convey Intelligence to Edge Aggregation Analytics 25
Natascha Harth, Kostas Delakouridis and Christos Anagnostopoulos

Publish-Subscribe Based Monitoring Model for Wireless Sensor Networks 45
Kemal Cagri Serdaroglu, Tevfik Kadioglu and Sebnem Baydere

Toward Data Governance in the Internet of Things 59
Sabrina Sicari, Alessandra Rizzardi, Cinzia Cappiello, Daniele Miorandi and Alberto Coen-Porisini

Advancing Cognitive Cities with the Web of Things 75
Sara D'Onofrio, Simone Franzelli and Edy Portmann

Transferring Wireless High Update Rate Supermedia Streams Over IoT 93
George Kokkonis, Kostas E. Psannis, Manos Roumeliotis, Yutaka Ishibashi, Byung-Gyu Kim and Anthony G. Constantinides

Smart Connectivity for Internet of Things (IoT) Applications 105
Albena Mihovska and Mahasweta Sarkar

A Blockchain-Based Storage System for Data Analytics in the Internet of Things 119
Quanqing Xu, Khin Mi Mi Aung, Yongqing Zhu and Khai Leong Yong

Ensuring Quality of Service in the Internet of Things 139
Giacomo Tanganelli, Carlo Vallati and Enzo Mingozzi

Wireless Sensor Networks-IoT Infrastructure 165
Azadeh Zamanifar

RFID-Based Multi-level Sensing Network for Industrial Internet of Things

S. Amendola, C. Occhiuzzi, S. Manzari and G. Marrocco

Abstract A wireless sensor network fully based on battery-less radio frequency identification (RFID) devices is here proposed for application to the emerging Industrial Internet of Things. The hierarchical structure of the network enables a multi-level monitoring of complex spaces hosting industrial equipments. A multi-antenna configuration permits to select the zones of the space to be monitored while custom RFID boards, capable to host several kinds of sensors, permit to capture both environmental parameters (e.g., temperature, humidity, and light) and the human interaction with things. The system provides the real-time detection of a plethora of complex events ranging from critical environmental accidents, up to the (un)authorized access to a critical area and the tampering/overloading of equipments. The potential of the proposed sensor network is finally demonstrated through an application to the monitoring of a real electrical secondary substation cabin.

Keywords Wireless sensor networks · RFID · Multi-level sensing · Multi-level monitoring · Industrial Internet of Things · Real-time monitoring

1 Introduction

The current development of the Internet of Things (IoT), beside opening innovative scenarios in connectivity, gaming, leisure, and domotics, is also fostering changes in modern manufacturing, energy, agriculture, transportation, and in many others industrial branches where the possibility to improve the interactions between human and machines may generate unprecedented technical and economic opportunities [1]. Gathering information about environments and processes will increase the

S. Amendola · C. Occhiuzzi (✉) · S. Manzari · G. Marrocco
RADIO6ENSE S.r.l, Rome, Italy
e-mail: amendola@info.uniroma2.it

S. Amendola · G. Marrocco
Pervasive Electromagnetic Lab, University of Roma Tor Vergata, Rome, Italy

R.R. Yager and J. Pascual Espada (eds.), *New Advances in the Internet of Things*,
Studies in Computational Intelligence 715, DOI 10.1007/978-3-319-58190-3_1

capabilities to control complex systems and to predict events, thus optimizing the production, the security, and the overall efficiency. This particular implementation of the IoT, denoted as *Industrial IoT*, is hence referred to wireless sensor networks which are characterized by a high level of autonomy and reconfigurability and above all by a minimum impact on costs, energy, and procedures. Sensors will be added to existing machinery without compromising their integrity at the purpose to incrementally upgrade their functionalities up to achieve, over the long term, a fully automated, flexible, networked, and data-oriented industry [2].

Energy-autonomous wireless sensors for application to ambient monitoring, personal tracking, cold chain, and manufacturing control have been greatly improved in the latest years mostly thanks to the well-assessed radio frequency identification (RFID) standard EPC C1G2 which can now offer sensing functionalities [3] besides the basic identification capability. This kind of sensors demands for a rather limited maintenance in comparison with the more assessed wireless technologies such as ZigBee, Bluetooth, or Wi-fi [4]. The required energy is indeed provided by external interrogators, which can interact with a multiplicity of sensors, thus enabling a single- to multi-point link with a remarkable reduction of the overall wiring. Small batteries and energy harvesters like solar panels can be used as well to support the low-power sensing activity while the communication is based on electromagnetic backscattering and can hence be considered as passive. The convergence between IOT and RFID tech is a well-known driving force toward the real implementation of what the authors defined "the last meters of the Internet of things," i.e., the physical layer of IOT systems [5–8].

The RFID energy-autonomous sensors which are nowadays available on the market, or which are being experimented worldwide in research laboratories, can be classified into two sets: (*i*) low-cost and qualitative *analog tags* for item-level applications and (*ii*) medium-cost electronic-packed *digital tags* which permit an accurate and versatile sampling of physical parameters. The first class, mainly comprising traditional RFID passive tags, exploits the interactions between tag antenna and environment to indirectly gather sensing information and hence can be affected by many uncertainties sources [9]. The latter family can instead include real and even COTS sensors[10–14] and are suitable to produce accurate data. Moreover, their cost is currently decreasing so that they are becoming attractive also for massive applications.

Although many examples of RFID-based sensors have been recently proposed by both academia and industry (refer to [3] for a review), the deployment of a fully autonomous wireless sensor network completely based on RFID technology is still in an embryonic stage. Some early successful examples are human activity monitoring [15, 16], ranging and localization of people and objects [17], bus fleet monitoring and scheduling [18], and workplace safety management [19].

To the best of our knowledge, this contribution introduces for the first time the complete design and implementation of an *RFID-based industrial sensor network* for application to critical infrastructures such as pipelines, smart grids, and power plants through the proper integration of the above two classes of analog and digital RFID sensors within the machineries and the nearby environment. The goal to be achieved is an autonomous and easily reconfigurable wireless sensor network, suitable to be employed in an industrial scenario with a minimum impact on the existing infrastructures. The architectures of both entire network and sensors are conceived to be modular and scalable, such as to include several sensing devices which can be easily repositioned into the environment. The paper addresses the design of both hardware and software components, and it is organized as follows: The multi-level architecture of the network is introduced in Sect. 2. A new topology of multi-function RFID sensing board is then described in Sect. 3, while Sect. 4 resumes the implementation of the control and coordination software. The deployment of the network in a real environment and some examples of multi-parametric monitoring are finally described in Sect. 5.

2 Architecture of the RFID Sensor Network

The proposed RFID sensor network (hereafter RFID-SN) is organized as a multi-level hierarchical architecture (Fig. 1) suitable to achieve spatial selectivity and sensing selectivity. From a logical side, the *space* under observation is partitioned into *M zones*. Each zone includes N_m *things* of interest, and the *nm*-th *thing* has K_{nm} *attributes* to be monitored along with the time. Finally, the proper processing of the K_{nm} attributes (singularly or combined) defines an *event*, i.e., any occurrence that is relevant for the industrial infrastructure. This scheme is physically implemented by considering a multi-channel interrogation module, i.e. the RFID reader, connected via coaxial cables to transmitting/receiving antennas (A_m),

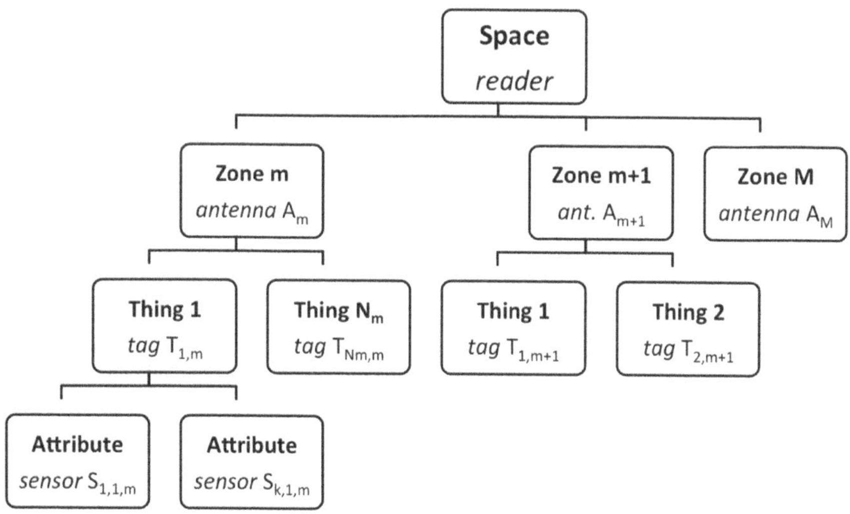

Fig. 1 Schematic multi-level architecture of the RFID sensor network

which interrogates the different RFID sensor tags (T_{nm}) properly dispersed into the environment. The electromagnetic coverage of each antenna within the space, which depends on the antenna gain, the radiated power, and the interaction of the electromagnetic field with the nearby environment [20], defines the extension of a zone. The RFID sensor tags, either analog or digital, identify the *things* to be monitored and may be integrated with one or more sensing mechanisms generating the measured data about the K_{nm} attributes of the thing itself. The number *C* of the independent channels (each channel being a sensor signal coming from a tag) produced by this architecture, e.g., the number of signals $\{S_{knm}\}$ collected by the reader node, is hence as follows:

$$C = \sum_{m=1}^{M} \sum_{n=1}^{N} K_{nm} \tag{1}$$

The reader node is managed by a control and command software living into the reader unit itself or into a remote system. Readers of different zones can be connected through Ethernet/Wi-fi links to a higher-level network whose description is, however, out of the scope of this paper.

This architecture offers a great freedom concerning the on-site physical reconfigurability (addition, repositioning, and dismantling of sensors) and, accordingly, in the space granularity of the surveillance. A remote dynamic handling of the power is moreover possible, i.e. the power emitted from the reader and even from the single interrogating antenna can be dynamically modified at the purpose to properly shape the extension of the zones and focus the available resources on the most critical areas where abnormal events have been detected.

The reader's antennas interact with all the kinds of tags by using the backscattering modulation principle: the energy needed by the sensors for data acquisition is directly scavenged from the electromagnetic field radiated by interrogation system, while no energy is wasted by the tag for the communication with the reader. The reader unit can interact with the sensors in time division according to a dynamic strategy allowing a periodic activation of all the beams to control the whole space volume, or, alternatively, trigger a reduced set of reader antennas at the purpose to control a subregion of the environment with a higher data rate.

2.1 Analog Signals

The analog tags (i.e. the conventional RFID tags) are displaced onto moving parts of the *space*, like cabinet windows or mobile equipments, as well as onto the access doors and on the wall of critical areas of the infrastructure to be monitored. The data produced by those tags is the level of the electromagnetic field (in the form of received signal strength indicator (RSSI)) they backscatter toward the A_m antenna of the reader during the interrogation. During an initial calibration, the system stores the electromagnetic fingerprint of the environment, i.e. the values of the RSSI produced by the various sensors in stationary conditions (as in the case of an anti-theft system). Accordingly, any geometrical change of the environment such as somebody moving inside [8], the interaction with doors, cabinets, or with any other critical device will produce an *environmental modulation* of the backscattered fields, and it will be perceived by the system as a perturbation of the RSSI collected by the reader.

Due to the uncertainties related to such measurements [9], these analog RFID tags are preferably used as threshold sensors since they provide information about those events that are characterized by a strong contrast between normal and abnormal working conditions. The intelligence to retrieve the sensed data is hence mostly concentrated at the reader side, which could be equipped with detection and classification algorithms (not addressed here) which are applied to raw RSSI data to recognize specific events (refer to [15, 21] for some idea of RFID-based pattern recognition). Typical threshold events in an industrial environment would be flooding, open/closing of doors and cabinets, and shadowing/scattering caused by human presence. A particular case of this class of sensors are RFID badges [22] of workers which can be recognized and identified by the system in critical areas, as better discussed later on by means of a real-life example.

It is worth noticing that recalibration of the event detection algorithm could be required in case of a severe modification of the environment close to the specific tag, like the placement of a new big object or the repositioning of a cabinet which may substantially alter the ambient backscattering of some tags. Accordingly, the pre-defined RSSI thresholds used to identify anomalous events have to be retuned. Instead, minor geometrical changes like the placement/motion/addition of a small thing far from the specific tag are expected to produce only a negligible effect especially if threshold detection is applied. Anyway, these static artifacts could be fully removed by recollecting the signal baseline following a remote request, or by an automatic algorithm.

2.2 Digital Signals

The digital tags are instead provided with specific internal/external sensors that produce quantitative data about the specific physical parameters under observation (light, humidity, temperature, deformation, radioactivity, and others). A digital tag has to be considered as a real multi-channel sensor node, even if the local computation capability is rather modest and restricted to sensor handling and configuration. Each tag hence reproduces the global multi-level structure that is implemented for the general architecture of the RFID-SN since the logic unit of the board can selectively activate one or more sensors according to the request coming from the reader. Data produced by the digital tags is directly suitable for a remote interpretation.

As a whole, the RFID-SN is hence capable to detect discrete events as well as to collect continuous variation of physical parameters by using a unitary infrastructure and a single communication protocol.

3 The Configurable RFID Sensing Breadboard

The core of the RFID-SN is a proprietary digital tag, hereafter denoted as *Radio-board*, that enables multi-channel battery-less sampling and transmission of environmental parameters.

The *Radio-board* is based on a new family of RFID chip transponders [10] providing a native integrated electronics for sensing activities beside the pure identification features. In particular, the selected IC includes an analog-to-digital converter (ADC) capable to control up to two analog external sensors and an integrated temperature sensor with programmable dynamic range in the interval –40/150 °C. This IC can be used in a fully passive mode (synchronous modality), i.e. the energy required for activation and actions is entirely harvested from the electromagnetic waves emitted by the remote interrogator, or in battery-assisted mode, i.e. a local battery can provide additional energy to improve the IC sensitivity (from −5 dBmW down to −15 dBmW) for extended read range and, above all, to perform periodic measurements even in the absence of the reader (asynchronous/data logging mode). The possible additional sensors which can be connected to the chip are any resistive, capacitive, or optical device, provided that its power consumption is compliant with low-power applications.

In order to master the wide range of functionalities the IC is provided with, the transponder element was engineered to make it operating in several radiation and sensing modalities while making use of a same mother PCB layout, thus speeding up the prototyping and customization of new products.

The *Radio-board* (Fig. 2) is logically composed of three parts:

(i) a radiating element made of a meander line antenna (MLA);
(ii) a spiral impedance transformer connected to the IC and to a tuning inductor L_T;
(iii) additional expansion traces for battery and sensors interconnections.

Both the MLA and the spiral traces are partly interrupted in several points (trace gaps M in the MLA and N in the spiral). Two further gaps (SW1 and SW2) split the transformer section from the MLA. The device can be globally regarded as a multi-port antenna. By properly selecting the tuning inductor and the subset of connected trace gaps, the distribution of the surface current over the antenna and, accordingly, the impedance and the gain, can be shaped for a specific application and positioning. For instance, acting onto the MLA, the size of the antenna is modified and hence also its gain and impedance, while by connecting some gaps, the spiral could be enlarged or reduced as needed for the specific

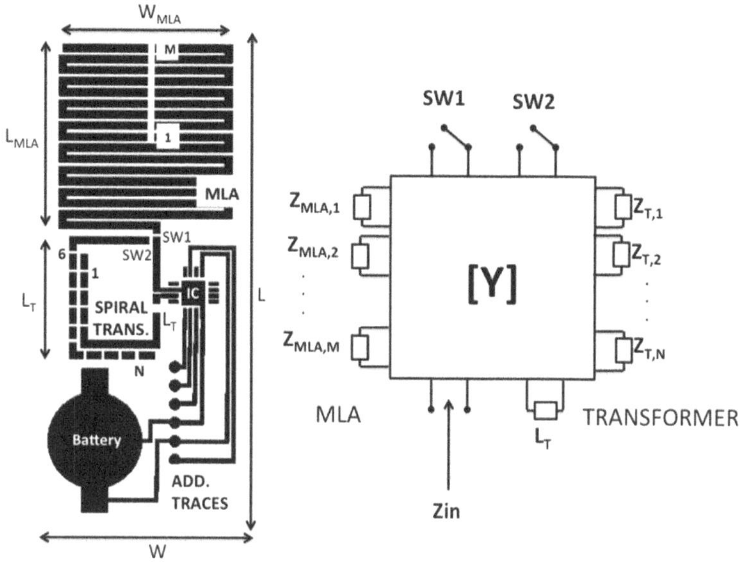

Fig. 2 (*Left*) Schematics of the customizable sensing breadboard architecture comprising the radiating meander line element (MLA), the spiral impedance transformer, and the expansion traces size in (mm): $W = 28$, $L = 66$, $W_{MLA} = 24$, $L_{MLA} = 26$, $W_T = 12.5$, $L_T = 20$ (mm); traces width: 1 mm (MLA and spiral) and 0.25 mm (sensor traces) (*Right*) A Multi-port model

microchip. Finally, depending on the status of the SW1/2 ports, the board can be used either as a tunable stand-alone RFID sensor board (SW1 closed and SW2 open) for application onto low permittivity and low losses materials, or as a basic module (SW1 open and SW2 closed) including the sensors, the chip, and the spiral transformer to be electromagnetically coupled to an external antenna, like a patch booster, for application over metals, or concrete walls.

Figure 3 shows a parametric exploration of the simulated realized gain [23] of the Radio-board in free space by acting either only on the trace gaps of the MLA or only on the spiral transformer ($Z_C = 31 - j330\ \Omega$). In the former case, replacing trace gaps with short circuits produces approximately a constant shift of about 10 MHz/gap. In the latter case, instead, the effect is less uniform but still suitable for a finer tuning of the impedance. By acting of the inductor it is finally possible to adjust the residual antenna reactance and hence to maximize the peak of the realized gain.

3.1 Application Examples and Performance

The potentiality of the proposed antenna architecture is here discussed by the help of two examples involving the full stand-alone breadboard radiating in air and the

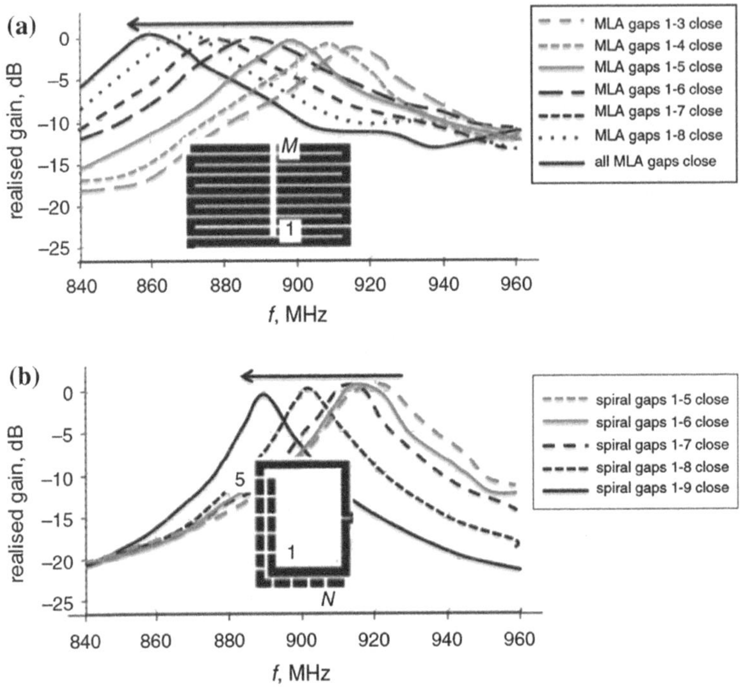

Fig. 3 Example of Radio-board tuning. Frequency shift of the broadside realized gain by connecting an increased number of the trace gaps of **a** the MLA and **b** of the spiral transformer length

same board placed onto an external patch booster for application over metals, or concrete walls.

The board prototype was fabricated by etching a 0.8-mm FR4 PCB. Flexible configurations on Kapton substrates can be manufactured as well.

In the first exercise, the SW1/2 gaps were configured so that the MLA is physically connected to the input section. Figure 4 shows the optimized realized gain close to 0 dB @ 868 MHz, having good agreement between measurements and simulations.

In the second example, the board was backed by a doubly folded patch [24], for improved operations over metal and lossy materials. In this case, the SW1-2 gaps were configured so that the MLA is physically disconnected from the input part. The spiral loop is hence inductively coupled with the radiating slot of the booster. The numerical multi-port characterization of the board over the patch included also an infinite ground plane where the tag is assumed to be attached on. As shown in Fig. 5, the device exhibits an appreciable realized gain (with/without battery) despite the close proximity of the ground plane. The presence of the battery yields worse performance in terms of realized gain, with respect to the battery-less

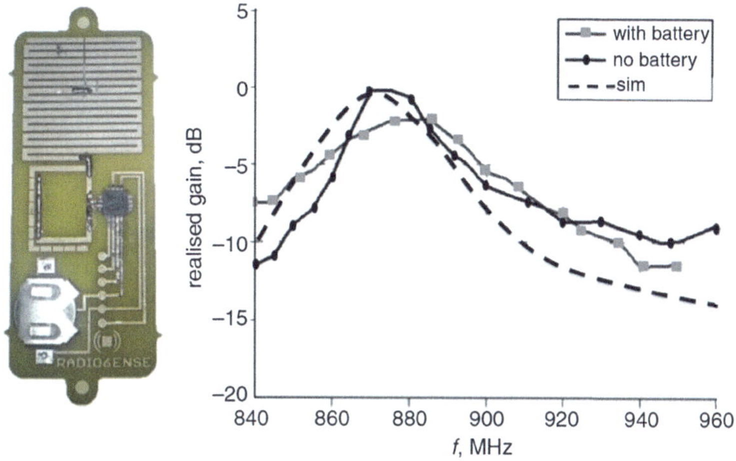

Fig. 4 Stand-alone RFID breadboard as optimized for working in air in UHF band. (*Left*) Prototype (optimized parameters $L_T = 47$ nH, SW_1 = close, SW2 = open, $Z_{T1,10} = \infty$, $Z_{T11\text{-}15} = 0$, $Z_{MLA1} = 0$, $Z_{MLA2\text{-}6} = \infty$); (*Right*) Simulated and measured realized gain. (with/without battery)

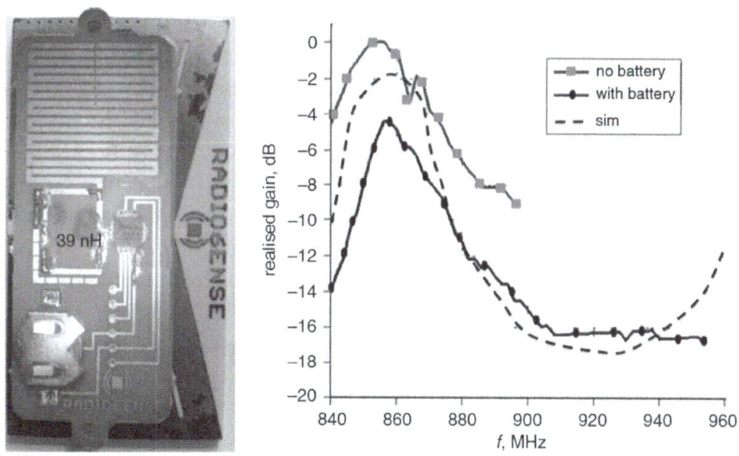

Fig. 5 Board backed by a patch booster (75 mm × 40 mm × 3 mm [24]) optimized for working on metal in UHF band. (*Left*) Prototype; (*Right*) Simulated and measured realized gain. (Optimized parameters $L_T = 39$ nH, SW_1 = open, SW2 = close, $Z_{T1,3} = 0$, $Z_{T3,10} = \infty$, $Z_{T11,15} = 0$, $Z_{MLA1,6} = \infty$)

configuration probably due to parasitic currents, flowing into the IC through the battery traces, which are enhanced by the presence of surrounding metal layers.

The experimented read ranges of the board at 868 MHz, when the reader emits 3.2 W EIRP, are resumed in Table 1 for the two possible modes of operations.

Table 1 Read range of the Radio-board in various configurations

	In air (m)	On metal/concrete with patch booster (m)
Battery-less	2	2.3
Battery-assisted	6.5	7

The digital tag has to be considered as a real multi-channel sensor node, even if the local computation capability is rather modest and restricted to sensor handling and configuration. Each tag reproduces at the lowest level the concept of a modularity own to the architecture of the RFID-SN since the logic unit of the board can selectively activate one or more sensors according to the custom commands that are elaborated by the control software running on the central unit and sent to IC through the A_m reader antenna.

4 Control and Coordination Software

The RFID-SN is governed by a software module (hereafter denoted as *RadioScan*), written in C# .NET, which implements the remote control over the multi-level hierarchical architecture of the network in Fig. 1. The structure of network is declared by means of an XML file which is associated to *RadioScan*. This file can be modified at run-time for achieving a dynamic control over the system, for example, to switch from asynchronous to synchronous mode and to increase the sampling time within a specific zone if an anomalous event was suspected.

The Config file is conceived to implement the tree diagram in Fig. 1. *RadioScan* sets the operative configuration of the sensor network by (*i*) selecting the zones of the space to be controlled (by switching on and off the corresponding antenna of the reader); (*ii*) defining the specific interrogation modalities of the zones in terms of sampling rate, frequency, and power emitted by each reader antenna; (*iii*) selecting the things of each zone to be monitored and the relative attributes (by enabling/disabling tags and activating the embedded sensors). Figure 6 shows an example of Config file describing a network consisting of four reader antennas sourced at 868 MHz by 31 dBm power according to the sequential rotation {1, 2, 4, 3}. The spatial architecture is defined in the `<NetworkConfiguration>` section where each antenna of the reader is associated with the list of the tags to be interrogated within the corresponding zone of the space (`ZONE 1` "`name=`" `tag_list_A.1`). Those Radio-board that are equipped with multiple sensors ("`type`" = `Radio-Board`) require additional fields for the selection of the on-board sensors and the settings of the corresponding sensor front end, such as the type of the sensor, the voltage levels defining the dynamic range and the resolution of the sensors, and the parameters for data logging functionalities.

The output of the software are (*i*) a log file containing the current network configuration (active reader antennas and corresponding detected sensors) which is

```
<InterrogationSettings>
        <add key="FrequencyRegion" value="European" />
        <add key="Frequency(MHz)" value="868" />
        <add key="Power(dBm)" value="31" />
        <add key="ReaderAntennas" value="1243" />
        <add key="Mode" value="RealTime" />
        <add key="SamplingTime(sec)" value="1" />
        <add key ="TCP/IP_stream" value = "true"/>
</ InterrogationSettings >

<NetworkConfiguration>

<ZONE1 name="tag_list_A.1">
   <tag name="T1" type="analog"></tag>
   <tag name="T2" type="RadioBoard

   Sensor Enabled ( RSSI="true" Temp="true"
                    Ext1="false" Ext2="true" Battery="false")

   Sensor Types   (Sensor1="-" Sensor2="Light" )

   Sensor Front-End Settings (V1="210" V2="310" ground="false"
                              Rref="8" current="31"....)

   DataLogger Settings (State="Start"  Interval="1"
                        Delay="6" Storage="normal"
                        Form="outoflimits"...)</tag>

   <tag name="T3" type="analog"></tag>
</ZONE1>

<ZONE2 name="tag_list_A.2">
   <tag name="T4" type="RadioBoard
   Sensor Enabled ( RSSI="true" Temp="true"
                    Ext1="false" Ext2="true" Battery="false")
</ZONE2>
```

Fig. 6 Example of Config file defining the configuration of the network

automatically saved at run-time when the software is started and (*ii*) a formatted string containing the time stamp and the (multi)sensor data of each tag at the current interrogation cycle. The string is both stored in a local text file and streamed over Ethernet port for remote processing.

In a possible complete architecture, these outputs could be accessed in real time by an upper decision layer (whose description is out of the scope of this paper) that implements detection algorithms [15] and, if needed, sends back the control software some input command to consequently update the network. Provided that each sensing node is reconfigurable via software (similarly to [25]), the RFID sensor network as a whole is definitely provided with the capability of self-configuration, which is a key requirement for IOT platforms. Moreover, direct tag-to-tag communication could be in principle possible by means of a pure backscattering modality, as demonstrated in [26, 27], thus fostering in the future an autonomous data exchange among nodes like in the more complex M2M devices.

5 Application to the Physical Security in Electric Plants

A valuable application of the described RFID-SN is the protection of critical infrastructures against cyber and physical attacks. Security, in its meaning of defense against cyber-attacks and threats, has been traditionally considered not to be a prominent issue for critical infrastructures such as pipelines, smart grids, and power plants, even though recent critical events demonstrated how deep is the relationship between cyber and physical worlds [28].

In the framework of European Horizon 2020, the project *Security in trusted SCADA and smart-grids* (SCISSORS, www.scissor-project.com) addresses the design of a holistic, multi-layered, security monitoring and mitigation framework, spanning all the issues related to the deployment of a critical infrastructure such as the control (*i*) of the environment, (*ii*) of the network traffic, (*iii*) of the hardware and software system components, (*iv*) of the people accessing the infrastructure, and (*v*) the independent monitoring of the control process itself. The environmental sensing and monitoring layer of SCISSORS are demanded to the proposed RFID-SN.

A first version of the RFID-SN was deployed and preliminary tested within the electrical transformer secondary substation of the University of Rome Tor Vergata (Fig. 7a). Similarly to other smart grid substations, the bunker room is located in the basement of the building and it is a restricted access area. The room contains two working transformers, several control cabinets, a couple of electric generators, and many high-power cable bundles.

5.1 Events to Be Detected

The events to be detected were the authorized/un-authorized accesses to the cabin, possible tampering of the machineries, humidity changes and flooding of sensitive areas, and power overload of wire harnesses. At this purpose, the Radio-board were equipped with humidity and light sensors and with high-temperature external probes. Analog tags were used as well to detect intrusions and mechanical changes of the room. Each event has been detected through the processing of a single attribute of one thing inside the space or through a combination of them (Table 2).

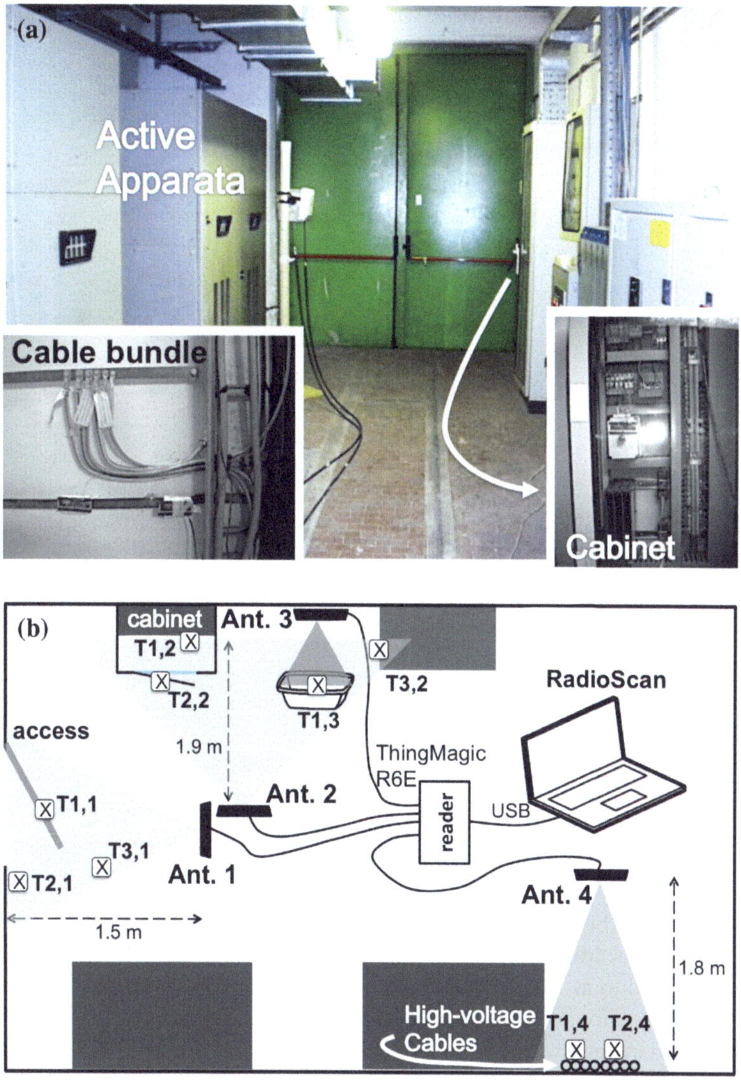

Fig. 7 **a** Electrical transformer secondary substation of the University of Rome "Tor Vergata." **b** Schematic representation of the deployed RFID-SN. The *gray* triangles highlight the read region (zone) of each antenna

5.2 Network Configuration

The configuration of the RFID-SN is sketched in Fig. 7b. A 1 W long-range RFID reader (ThingMagic M6E [29]) connected to four antennas (circularly and linearly polarized patches) was used to monitor four different zones ($M = 4$) in the cabin: access (A_1); cabinets and energy meters (A_2); flooding sensitive area (A_3);

Table 2 Events to be detected by the RFID-SN inside the electrical cabin and corresponding attributes to be measured

Event	Attributes
Authorized/un-authorized access	RSSI from tags installed onto the access door
	ID code from wearable tags
	Light on/off
Flooding	RSSI from tags placed over the floor
	Ambient humidity
Harness overload	Temperature variation of cables
	And bundles of cables
	(High and low range)
Manumission of cabinets	RSSI from tags placed onto the cabinet window
	Temperature variation inside the cabin

Table 3 Network configuration

Antennas (zones and events)	Tags	Sensor channel
A_1: Zone 1 (Authorized/un-authorized access)	$T_{1,1}$—W-tag	$S_{1,1,1}$: RSSI
	$T_{2,1}$—Radio-board	$S_{1,2,1}$: RSSI
		$S_{2,2,1}$: Light (S133-14 p.diode)
	$T_{3,1}$—W-tag	$S_{1,3,1}$: RSSI
A_2: Zone 2 (Manumission of cabinets and flooding)	$T_{1,2}$—Radio-board	$S_{1,1,2}$: Temp (internal sensor)
	$T_{2,2}$—W-tag	$S_{1,2,2}$: RSSI
	$T_{3,2}$—Radio-board	$S_{1,3,2}$: RH% (HCZ-D5 sensor)
A_3 : Zone 3 (Flooding)	$T_{1,3}$—W-tag	$S_{1,1,3}$: RSSI
A_4 : Zone 4 (Harness overload)	$T_{1,4}$—Radio-board	$S_{1,1,4}$: Temp (PT1000)
	$T_{2,4}$—Radio-board	$S_{1,2,4}$: Temp (internal sensor)

high-power cable bundles (A_4). The set of RFID tag (Table 3) comprised five Radio-board, embedding heterogeneous sensors, and four analog sensor tags [22], hereafter referred to as W-tags. The latter are platform-tolerant tag that can be used as wearable badge for automatic access identification of operators, as RSSI markers over doors and cabinet windows to detect a possible interaction with a persons and even deployed over the ground and wall for flooding control. The overall number of channels of the network was $C = 10$.

Measurements were carried out in both rest (stationary) and operative (dynamic) conditions. The critical events listed in Table 2 were emulated several times by the help of volunteers.

5.3 Flooding and Humidity

Flooding is a recurring event in smart grids, especially when the infrastructure comprises several underground cabins. In the case of partial flooding, the signals backscattered by the W-tags placed on the critical regions of the floor are strongly perturbed. Eventually, when a tag is sensibly submerged by water, it becomes undetectable by the reader due to the abrupt change of the electromagnetic parameters of the surrounding medium which detunes the antenna. In addition to this threshold detection, the Radio-board with relative humidity sensors can be used to detect abnormal variation of the environmental relative humidity (%RH) which can be related to flooding.

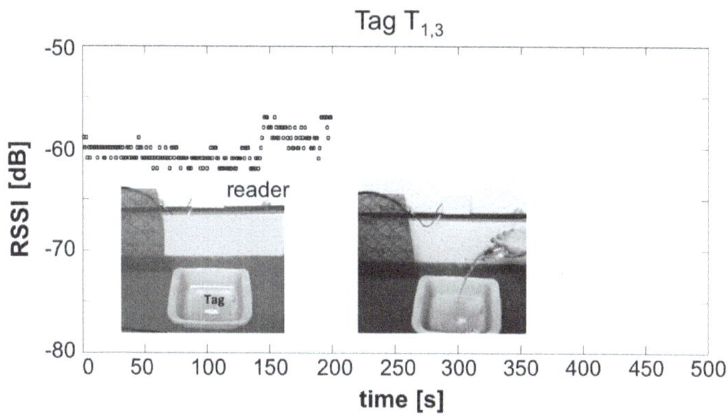

Fig. 8 Backscattered power from the analog tag $T_{1,3}$ during the simulation of a flooding event

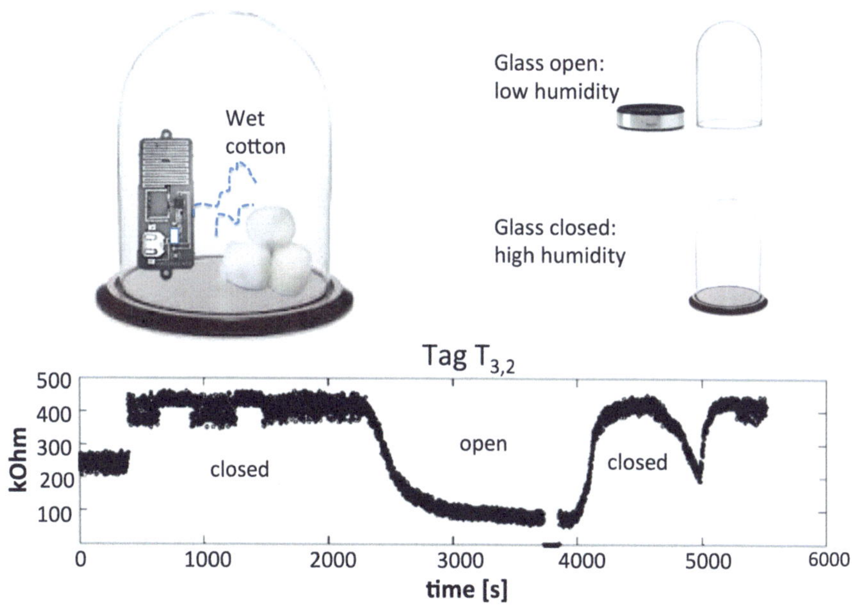

Fig. 9 Impedance of the humidity sensor connected to the Radio-board $T_{3,2}$ measured during cyclic humidity variations induced by opening/closing the glass bell

The flooding event was emulated (Fig. 8) by filling a plastic basin with water. A W-tag ($T_{1,3}$) was placed on the bottom of the basin. In normal conditions (absence of water), the tag was detected by the antenna A_3 with a stable level of RSSI ($S_{1,1,3}$). Then, as soon as it was covered by the liquid, it was severely mismatched and became unreadable by the antenna.

The humidity change was instead simulated by placing the Radio-board $T_{3,2}$ equipped with the humidity sensor into a glass bell together with a piece of wet cotton. When closing the glass, the internal relative humidity gradually increased up to saturation. Then, it rapidly came back to the initial ambient condition as soon as the top was removed. An example with some open/close cycles is reported in Fig. 9 showing the impedance of the sensor ($S_{3,2}$) which is inversely proportional to the humidity level detected by the antenna A_2.

5.4 Harness Overloading

An anomalous working load of the cabin transformer could produce high currents over the distribution cables. Radio-board including internal temperature sensors and/or connected to external high-temperature probes can be placed over the cable harness to monitor their surface temperature which is related to the currents flowing into the cables themselves. Those sensors can be also used to obtain indirect information about the aging of the dielectric insulators of the harness.

In the present experiment, some events of power overloading were reproduced by manually warming up two harnesses inside the cabin by using a heat gun. Figure 10 shows the temperatures ($S_{1,1,4}$ and $S_{1,2,4}$) detected by the two Radio-board ($T_{1,4}$ and $T_{2,4}$) attached over the two considered cables running along the perimeter wall of the cabin, the first one integrating a platinum thermo-resistance (PT1000) whose extremal sensitive part was at direct contact with electric cables and the second one detecting the temperature by its internal sensor.

5.5 Cabin Access and Manumission

In normal conditions, the door of the cabin is closed and the internal scenario is completely dark: Accordingly, a low-light signal can be collected by the light sensor of a Radio-board $T_{2,1}$ placed in proximity of the access doors. If somebody opens/tampers the door and gets inside the cabin, the system will detect an increase of the light (coming from outside or emitted by a torch or by any other light source), as well as a distortion of the RF fingerprint of the cabin due to the presence of moving people which perturbs the electromagnetic field produced by the reader antennas. Finally, if the subject is provided with an RFID badge, the passive

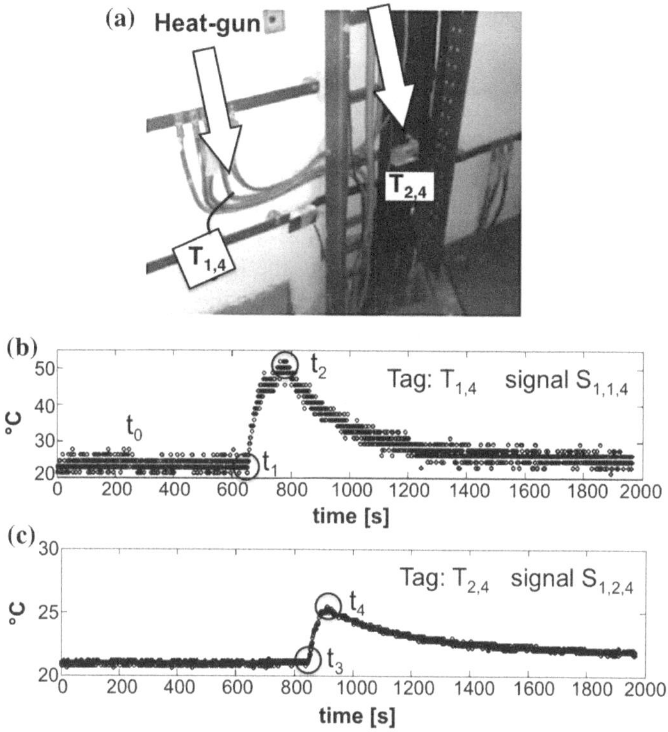

Fig. 10 Power overloading of two cables **a** equipped with Radio-board $T_{2,4}$ (internal temperature sensor) and $T_{1,4}$ (external PT1000 temperature sensor). **b** and **c** Temperature recording during an artificial warming by a heat gun in the time intervals (t_1, t_2) and (t_3, t_4) for the two cables, respectively

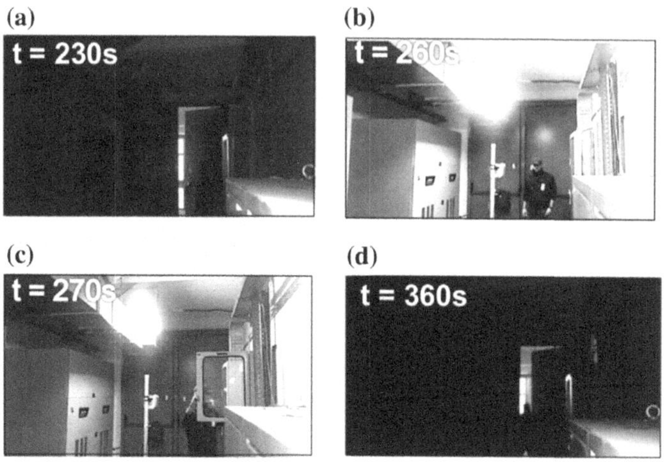

Fig. 11 Screenshots from the authorized access to the electric cabin and interaction with an equipment. **a** door opening; **b** light on and RF badge detection; **c** cabinet opening; **d** light off, badge recognized, door open and closed

network is able to verify his identity and his right to access the cabin, for instance, in case he is maintenance personnel.

5.5.1 Authorized Access

Figure 12 shows a subset of the signals recorded by the sensor network when an authorized technician came into the cabin for ordinary maintenance (screenshots in Fig. 11). In the initial reference condition, the light in the room was off ($S_{2,2,1}$ signal of Radio-board $T_{2,1}$) and the W-tags for the access control ($T_{1,1}$) and cabinet opening ($T_{2,2}$ returned stable RSSI values. No authorized people were detected inside the ambient (null signal from wearable tag $T_{3,1}$). The evident drop of the RSSI collected by sensor $T_{1,1}$ reveals the opening of access door.

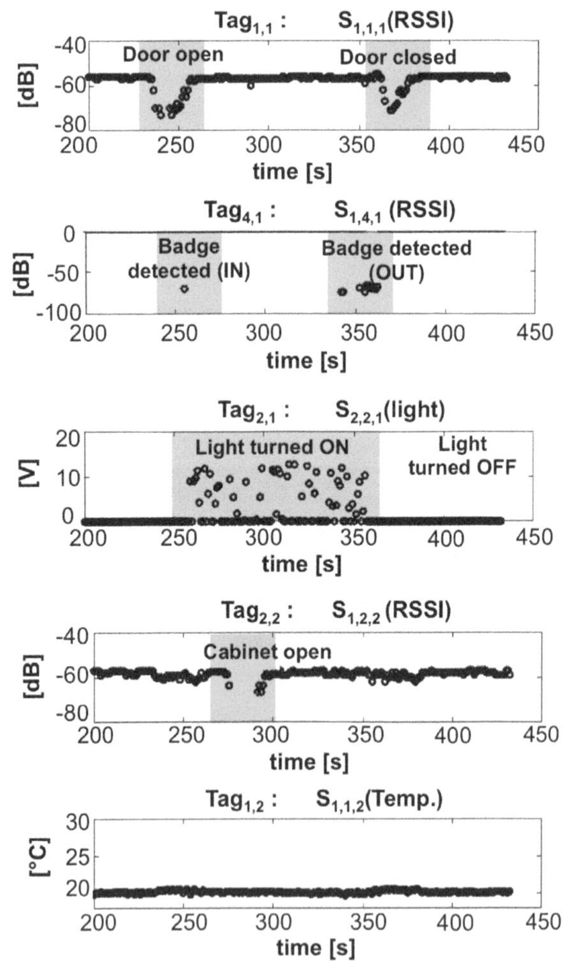

Fig. 12 Subset of signals collected by the RFID-SN in case of authorized access to the electric cabinet as sketched in Fig. 11

Immediately after, the person entering the room was automatically recognized by the system and classified as “authorized person” through his badge identification ($S_{1,3,1} \neq 0$). The maintenance technician turned on the light ($S_{2,2,1}$ switches to ON state) and opened the electrical cabinet (sensor on the door cabinet $T_{2,2}$ was no longer read in the open position) to perform ordinary operations, with no modification of the equipment temperature. Finally, the technician approached the exit door and turned off the light; the system detected again his badge and recorded the exit (Fig. 12).

5.5.2 Un-authorized Access and Attack

In a second experiment, the person entering the room was an intruder, i.e. he did not wear any RF badge, and he walked in the dark by using a torch, opened the windows of the cabinet, and artificially increased the temperature of an internal equipment to emulate a tampering event (Fig. 13). The multi-parameter recording by the RFID-SN is shown in Fig. 14 and could be interpreted, a posteriori, as follows: when the person came inside, the system recognized the door opening through a perturbation of the RSSI from the tag $T_{1,1}$. Since no pre-registered ID

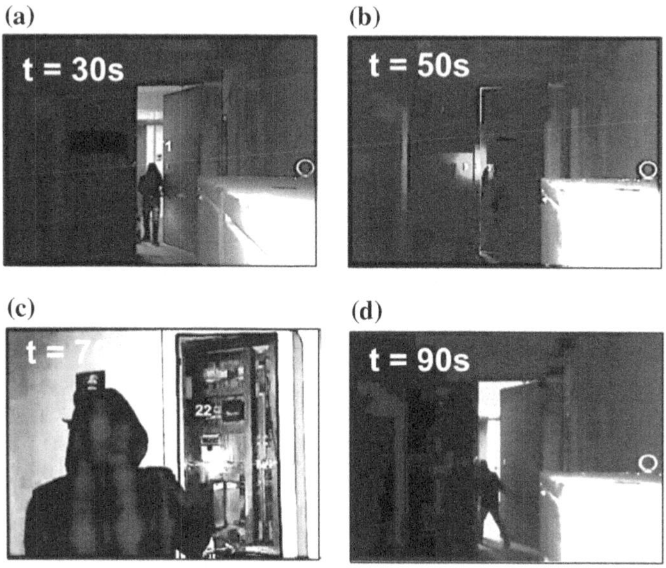

Fig. 13 Screenshots from the un-authorized access to the electric cabin with tampering. **a** door open; **b** torch pointing toward the light sensor; **c** cabinet opening and tampering; **d** attacker going outside

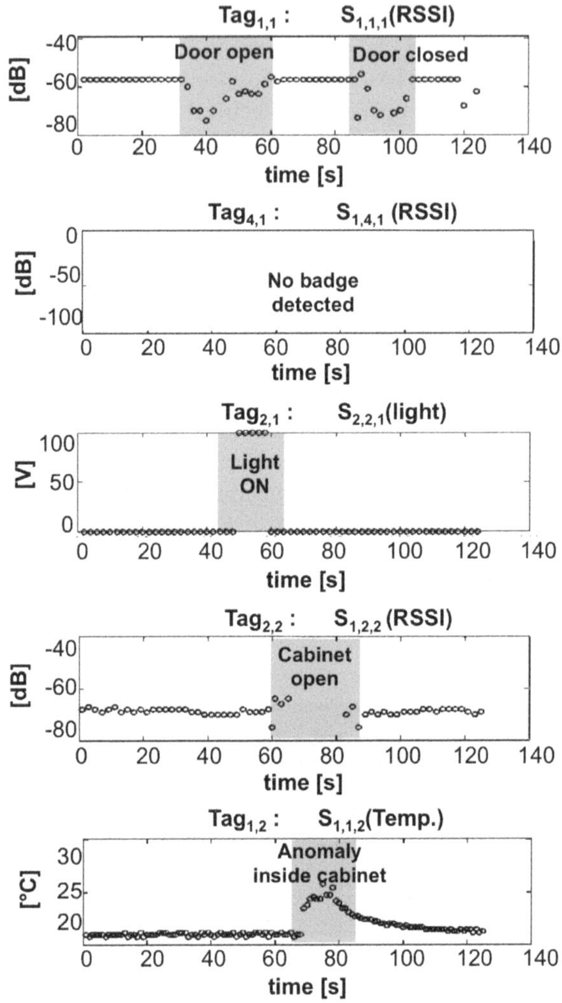

Fig. 14 RFID-SN measurements in case of an un-authorized access to the electric cabinet producing a thermal anomaly

code was detected, the person could be classified as an intruder. Then, a variation in the light level was revealed by the sensor $T_{2,1}$ for a short period suggesting that the intruder turned on the light just for a few seconds or used a torch. The sensor $T_{2,2}$ detected an interaction with the cabinet, and during this time interval, the internal temperature of the cabinet abnormally increased (sensor $S_{1,1,2}$ of $T_{1,2}$). This event could be considered as a warning of a potential power overload of some internal circuitry produced by a possible manumission. The sensor $T_{1,1}$ at the main door detected again an interaction when the attacker came out the cabin.

6 Summary and Conclusion

The proposed monitoring platform exploits the combined processing of analog and digital signals to detect anomalous event.

The hierarchical architecture enables a flexible and easily reconfigurable monitoring of a complex space as well as it permits to capture the user's interaction with specific nearby objects.

A custom transponder supporting multi-purpose sensing and radiation modes was specifically designed to provide the same layout with post-fabrication configurability by manually soldering tuning elements and wirelessly programming the IC logical unit.

An important issue is how the system complexity scales with respect to the size of the space under observation. The number of interrogating antennas, and hence of the cables, increases only linearly with the volume of the space to monitor, while it is independent of the number of things in each zone. Most of industrial-oriented RFID readers are provided with multiple antenna ports (up to four, as in the given example), and an even larger number of antennas could be addressed by using an electronic-controlled switch so that large spaces could be monitored with a unique centralized node.

The deployment of the network in the experiment required a try and error effort to identify the best position of the reader antennas so that all the tags were correctly read by the network. This procedure could, however, be driven by electromagnetic modeling, which includes the scattering of the nearby environment, and by evolutive optimization algorithms as in [30] for automatic antenna placement.

The proposed solution could find successful application to the empowering of SCADA (supervisory control and data acquisition) and video surveillance systems which are currently used in industrial infrastructures, thus producing both complementary and backup data.

In the framework of the SCISSOR project, a realistic test bed is currently running in an operational smart grid in Favignana Island (Italy) where the whole RFID-SN was successfully and permanently installed in September 2016. Figure 15 shows a snapshot of the dashboard that is remotely accessible from anywhere for the real-time visualization of the acquired data (see https://www.youtube.com/channel/UCkJHWrnq9bBJhUyQrJfwBlA for demo video (Fig. 15).

Furthermore, unlike the more conventional wired/wireless equipments for environmental monitoring and access control that sensibly suffer from the lack of a unique infrastructure [31], the proposed sensor network relays onto a well widespread and standardized protocol and on a growing set of COTS devices with clear benefits for the interoperability among services and the integration with existing industrial infrastructures. The system is hence suitable to be easily tailored and customized for combined access control, environmental as well as thing-level monitoring with minimal installation, maintenance, and dismantling times and costs (see Table 4 for a qualitative comparison).

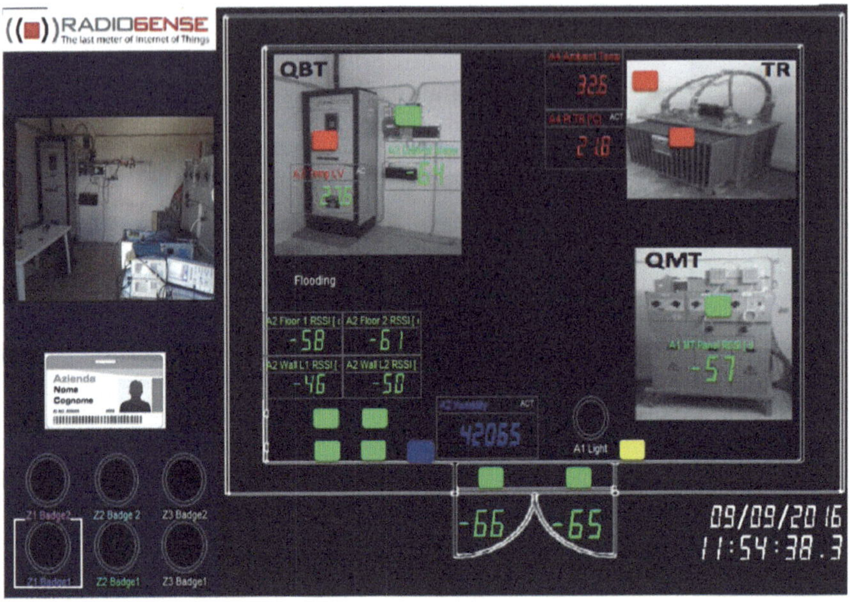

Fig. 15 Dashboard for the remote control of the RFID-SN installed within a smart grid in Favignana Island

Table 4 Industrial IoT Technologies

		RFID Sensors	Wired Sensors	Wireless Sensors
Costs	Installation	LOW	HIGH	LOW
	Maintenance	LOW	LOW	HIGH
	Power	LOW	LOW	HIGH
	Hardware	LOW	HIGH	HIGH
Benefits	Security	HIGH	HIGH	MEDIUM
	Scalability	HIGH	LOW	HIGH
	Reconfigurability			
	Interoperability	HIGH	LOW	LOW
	Sensor accuracy	MEDIUM	HIGH	HIGH

Finally, thanks to their local poor or even null computational capabilities, the RFID sensor nodes are expected not to be exposed to external cyber-attacks so that the whole security care could be entirely devoted to the reader node only.

Acknowledgements The work was supported by SCISSOR ICT project no. 644425, funded by the European Commissions Information and Communication Technology H2020 Framework Program.

References

1. Industrial internet of things: Unleashing the potential of connected products and services. World Econ. Forum Tech. Rep. (2015)
2. E. Brynjolfsson, A. McAfee, The Second Machine Age: Work, Progress, and Prosperity in a Time of Brilliant Technologies (W.W. Norton and Company, 2014)
3. C. Occhiuzzi, S. Caizzone, G. Marrocco, Passive uhf rfid antennas for sensing applications: Principles, methods, and classifications. Antennas Propag. Mag. IEEE **55**(6), 14–34 (2013)
4. W. Dargie, C. Poellabauer, Fundamentals of wireless sensor networks: theory and practice (Wiley, 2010)
5. G. Marrocco. et al., Rfid iot: a synergic pair. IEEE RFID Virtual J. **8** (2015)
6. M.A. Razzaque, M. Milojevic-Jevric, A. Palade, S. Clarke, Middleware for internet of things: a survey. IEEE Internet Things J. **3**(1), 70–95 (2016)
7. L. Catarinucci, D. De Donno, L. Mainetti, L. Palano, L. Patrono, M.L. Stefanizzi, L. Tarricone, An iot-aware architecture for smart healthcare systems. IEEE Internet Things J. **2** (6), 515–526 (2015)
8. S. Amendola, R. Lodato, S. Manzari, C. Occhiuzzi, G. Marrocco, Rfid technology for iot-based personal healthcare in smart spaces. IEEE Internet Things J. **1**(2), 144–152 (2014)
9. C. Occhiuzzi, G. Marrocco, Precision and accuracy in uhf-rfid power measurements for passive sensing. IEEE Sens. J. (99), 1–1 (2016)
10. SL900A, http://ams.com/eng/Products/UHFRFID/UHF-Interface-and-Sensor-Tag/SL900A
11. L. Catarinucci, R. Colella, L. Tarricone, A cost-effective uhf rfid tag for transmission of generic sensor data in wireless sensor networks. IEEE Trans. Microw. Theory Tech. **57**(5), 1291–1296 (2009)
12. A. Sample, D. Yeager, P. Powledge, J. Smith, Design of a passively-powered, programmable sensing platform for uhf rfid systems, in IEEE International Conference on RFID, Mar 2007, pp. 149–156
13. EM 4325, www.emmicroelectronic.com
14. http://www.farsens.com
15. C. Occhiuzzi, C. Vallese, S. Amendola, S. Manzari, G. Marrocco, Night-care: A passive rfid system for remote monitoring and control of overnight living environment. Procedia Comput. Sci. **32**, 190–197 (2014)
16. M. Buettner, R. Prasad, M. Philipose, D. Wetherall, Recognizing daily activities with rfid-based sensors, in Proceedings of the 11th International Conference on Ubiquitous Computing, ser. UbiComp '09. (ACM, New York, NY, USA, 2009), pp. 51–60. doi:10.1145/1620545.1620553
17. A. Costanzo, D. Masotti, T. Ussmueller, R. Weigel, Tag, you're it: Ranging and finding via rfid technology. IEEE Microw. Mag. **14**(5), 36–46 (2013)
18. W. Sriborrirux, P. Danklang, N. Indra-Payoong, The design of rfid sensor network for bus fleet monitoring, in 8th International Conference on ITS Telecommunications, 2008. ITST 2008, Oct. 2008, pp. 103–107
19. M. Sole, C. Musu, F. Boi, D. Giusto, V. Popescu, Rfid sensor network for workplace safety management, in 2013 IEEE 18th Conference on Emerging Technologies Factory Automation (ETFA), Sept 2013, pp. 1–4
20. G. Marrocco, E. Di Giampaolo, R. Aliberti, Estimation of uhf rfid reading regions in real environments. Antennas Propag. Mag. IEEE **51**(6), 44–57 (2009)
21. S. Amendola, L. Bianchi, G. Marrocco, Movement detection of human body segments: passive radio-frequency identification and machine-learning technologies. IEEE Antennas Propag. Mag. **57**(3), 23–37 (2015)
22. S. Manzari, S. Pettinari, G. Marrocco, Miniaturized wearable uhf rfid tag with tuning capability. Electron. Lett. **48**(21), 1325–1326 (2012)

23. F. Amato, G. Marrocco, *Self-Sensing Passive RFID: from Theory to Tag Design—an Experimentation* (European Microwave Conference, Roma, Italy, 2009)
24. S. Manzari, G. Marrocco, Modeling and applications of a chemical-loaded UHF RFID sensing antenna with tuning capability. IEEE Trans. Antennas Propag. **62**(1), 94–101 (2014)
25. M.S. Khan, M.S. Islam, H. Deng, Design of a reconfigurable rfid sensing tag as a generic sensing platform toward the future internet of things. IEEE Internet Things J. **1**(4), 300–310 (2014)
26. G. Marrocco, S. Caizzone, Electromagnetic models for passive tag-to-tag communications. IEEE Trans. Antennas Propag. **60**(11), 5381–5389 (2012)
27. P.V. Nikitin, S. Ramamurthy, R. Martinez, K.V.S. Rao, Passive tag-to-tag communication, in 2012 IEEE International Conference on RFID (RFID), Apr 2012, pp. 177–184
28. M.B. Kelley, The stuxnet attack on iran's nuclear plant was 'far more dangerous' than previously thought, Businessinsider.com. Tech. Rep. (2013)
29. http://www.thingmagic.com/index.php/fixed-rfidreaders/mercury6
30. E.D. Giampaolo, F. Fornì, G. Marrocco, RFid-network planning by particle swarm optimization. Aces J. **25**(3), pp. 263–272 (2010)
31. O. Monnier, E. Zigman, A. Hammer, Understanding wireless connectivity in the industrial iot, Texas Instruments, Tech. Rep. (2015)

Convey Intelligence to Edge Aggregation Analytics

Natascha Harth, Kostas Delakouridis and Christos Anagnostopoulos

Abstract In Internet of Things (IoT) environments, networks of sensors, actuators, and computing devices are responsible to locally process contextual data, reason and collaboratively support aggregation analytics tasks. We rest on the edge computing paradigm where pushing processing and inference to the edge of the IoT network allows the complexity of analytics to be distributed into many smaller and more manageable pieces and to be physically located at the source of the contextual information it needs to work on. This enables a huge amount of rich contextual data to be processed in real time that would be prohibitively complex and costly to deliver on a traditional centralized cloud/back-end processing system. We propose a lightweight, distributed, predictive intelligence mechanism that supports communication efficient aggregation analytics within the edge network. Our idea is based on the capability of the edge nodes to perform sensing and locally determine (through prediction) whether to disseminate contextual data in the edge network or to locally re-construct undelivered contextual data in light of minimizing the required communication interaction at the expense of accurate analytics tasks. Based on this decision making, we eliminate data transfer at the edge of the network, thus saving network resources for sensing and receiving data, by exploiting the nature of the captured contextual data. We provide comprehensive experimental evaluation of the proposed mechanism over a real contextual dataset and show the benefits stemmed from its adoption in edge computing environments.

Keywords Edge analytics · Aggregation analytics · Predictive intelligence · Communication efficiency · Context prediction

---

N. Harth · C. Anagnostopoulos (✉)
School of Computing Science, University of Glasgow, Glasgow G12 8QQ, UK
e-mail: christos.anagnostopoulos@glasgow.ac.uk

N. Harth
e-mail: n.harth.1@research.gla.ac.uk

K. Delakouridis
R&D Repado, B. Georgiou & Souri, Athens, Greece
e-mail: kostas.delakouridis@repado.com

R.R. Yager and J. Pascual Espada (eds.), *New Advances in the Internet of Things*, Studies in Computational Intelligence 715, DOI 10.1007/978-3-319-58190-3_2

1 Introduction

Edge analytics is an approach to efficient contextual data collection and analysis in which computation is performed on sensing devices (sensors, actuators), network switches or other devices (concentrators) instead of transmitting the whole data to a centralized computing environment, e.g., the cloud. Edge analytics has gained attention as the Internet of Things (IoT) paradigm of connected *things* (e.g., sensors, actuators, controllers, concentrators) has become more prevalent [1]. The *edge* of the IoT is where the action is. In IoT environments, high data rate sensors, e.g., video camera, environmental sensors, smart meters are becoming ubiquitous. Today, most high data volumes obtained from such sensors are stored *close* to the point of capture, and only a few are transferred to the cloud. In future, sending all the data from billions of IoT devices to the cloud can overwhelm the existing infrastructure. To overcome these issues, Edge Computing (EC) [2, 3] is emerging bringing contextual data processing, networking, and analytics closer to the IoT devices and applications.

EC represents a shift in which *intelligence* is pushed from the cloud to the edge, localizing certain kinds of analysis, e.g., aggregation operators over data streams and local decision-making [4]. This enables quicker response times, unencumbered by network latency, as well as reduced traffic, by *intelligently processing and relaying the appropriate analyzed data to the cloud.* A primary benefit of edge analytics is scalability. Pushing analytics algorithms to IoT devices alleviates the processing strain on enterprise data management and analytics systems, even as the number of connected devices being deployed by organizations—and the amount of data being generated and collected—increases [5]. By 2020, its projected there will be anywhere from 25 to 50 billion *things* connected to the IoT, i.e., up to seven connected things for every person on Earth. We can anticipate that these billions of connected things will generate data volume far in excess of what can easily be processed and analyzed in the cloud dealing with energy constraints (network lifetime), limited bandwidth, and network latency [6]. Unlike the cloud infrastructure, edge network exhibits the following properties: (i) heterogeneous hardware, (ii) unreliable low bandwidth communication network, (iii) limited, on-board energy budget and limited processing power. Moreover, edge analytics algorithms should not rely on any central coordinator and must be fault-tolerant as node/link failures are a common occurrence. Transferring large volume of data to cloud using low-power radio is often unfeasible due to energy and bandwidth limitation.

We envisage the *edge* of a network as a site for off-loading bandwidth and energy hungry sensor data. To generate value out of the large volume of data on the edge, we need energy-efficient, communication-efficient, autonomous, and lightweight contextual information processing algorithms. We abstract an edge network architecture through *edge nodes* forming a layer between *sensing/actuator nodes* and the cloud. Several Sensing and Actuator Nodes (SAN) are connected to each Edge Node (EN), e.g., cloudlet, sink node, powerful smartphone. Since ENs are located close to the SANs, contextual data should be intelligently transferred to them in real-time and in an energy efficient manner. Each SAN performs measurements and locally determines whether to transfer these measurements to the ENs or not in light of minimizing the required communication interaction (overhead) at the expense of accurate analytics tasks performed on the ENs. Based on this context, our idea rests on locally predict *whether* to disseminate sensed data or not within an edge network to **achieve quality analytics by being energy efficient**. In other words, we attempt to eliminate data transfer at the edge of the network, thus saving network resources for sensing and receiving data, by exploiting the nature of the captured contextual data. However, this comes at the expense of the quality of analytics tasks.

The fundamental requirement to materialize such predictive intelligence at the edge network is (i) the autonomous nature of SANs to locally perform sensing and disseminate data under analytics quality-driven rules and (ii) the capability of the ENs to locally perform lightweight and robust analytics tasks over the data received from their connected SANs. Our major goal is to examine the impact of this predictive intelligence on the quality of the aggregation analytics tasks on the ENs in light of extending the edge network lifetime and being communication efficient.

2 Overview and Motivation

2.1 Aggregation Analytics

All pieces of context captured by SANs and ENs in IoT environments, in general, **contextual information** sources are considered as continuous data streams, where analytics tasks are applied to extract statistical dependencies, aggregate analytics tasks, and infer new knowledge. Context-aware applications, crowd-sensing applications in IoT [7, 8], environmental and geophysical monitoring [9], e.g., forest monitoring [10–13] (through unnamed vehicles), agriculture monitoring [14], road traffic monitoring, surveillance, video analytics [15], marine environment monitoring [16], watershed monitoring systems [17, 18] and statistical analytics applications over large-scale data streams require efficient, accurate and timely data analysis in order to facilitate (near) real-time decision-making, data stream mining, and situational context awareness in IoT environments. The IoT computing devices revolutionize sensing in a wide range of application domains because of their reliability, accuracy, flexibility, cost effectiveness and ease of deployment. A contextual data stream contains values from contextual parameters corresponding to sources (e.g., humidity and temperature sensors of a smart-phones). A set of sources (e.g., mobile sensing and computing devices) captures contextual information representing e.g., a geographically monitored area or the condition of a road network. Context-aware and IoT applications exploit all such context, for instance, to (i) obtain the most affected areas by a fire in the last twenty minutes, (ii) identify a concept drift in certain parameters from eleven o'clock in the morning up to now by applying aggregation functions overs contextual values (e.g., `SUM`, `AVG`), (iii) infer the top-k recent congested segments of city road networks, or (iv) obtain regularly the highest pollution level within a time horizon in a smart city. Many critical IoT applications have been developed on top of contextual data streams captured by IoT devices for events identification. Events are related to critical aspects, e.g., security issues or violations of predefined constraints. For instance, in security and environmental monitoring applications, a monitoring infrastructure is imperative to apply an efficient mechanism to derive alerts when criteria are satisfied [19].

2.2 Overview and Literature Review

A baseline approach for materializing analytics tasks on the cloud is simply all IoT devices to transmit the contextual data from all sensing nodes to certain sink nodes, or back-end system. This has been realized in many previous studies [20–22]. In this case, analytics tasks are carried out by the back-end system on the cloud only, and not by the SANs or ENs at the edge of the network, despite their increasing computing capacity. Evidently, this solution, while practical, has many disadvantages, such as a high energy consumption incurred by transmitting the raw data to the cloud, the need for wireless link bandwidth, and high latency [3].

In the era of EC, instead, the desiderata are:

- *push* the analytics tasks close to the contextual data sources, i.e., to the ENs and
- *push* intelligence to SANs and ENs to *collaboratively* support edge analytics. ENs have to intelligently communicate with the SANs in an energy-efficient way, since communication efficiency is crucial to the prolonged lifetime of the edge network to support edge analytics.

We have distinguished two basic methods for edge analytics. The first method is based on the observation that the SANs and ENs, capable of local computation and sensing, create the possibility of analyzing and building (training) analytics models in a distributed way. In this class of edge analytics, e.g., [23–25], contextual data and/or model's meta-data are circulated within the edge network, which evidently requires energy for data and meta-data dissemination adding extra communication overhead. The second method refers to a group-based communication and single localized computation/processing scheme e.g., [13, 20, 26–30]. In this approach, an EN is responsible for a group of SANs and maintains a set of historical contextual data of each SAN within the group. Such localized method is communication efficient due to the reduced length of routing path from SANs to the cloud. To support such type of edge analytics, energy is consumed on communication, i.e., sending and receiving data from SANs to the EN, and computation, i.e., ENs are processing local data. However, since the cost of local processing and analytics tasks is nontrivial, we should take into account the trade-off between intra-edge-network communication and localized computation [31].

Both above mentioned basic methods are required to be efficient to support edge analytics in terms of computation and communication. The computational efficiency of the analytics tasks is a challenging research area, where recently distributed and large-scale statistical and machine learning algorithms emerge, e.g., [32]; this is beyond the scope of this paper. In the edge network communication aspect, we elaborate on the mechanism of the *selective data delivery*, which is adopted in many distributed computing and sensors environments [31]. We argue that this mechanism can be adopted on EC environments and support communication-efficient edge analytics. Specifically, this mechanism relies on the principle of bounded-loss approximation: a network node locally decides on delivering its sensed data to a network node based on local predictions of representative contextual data, which approximate the actual data given an approximation error bound. The intuition behind this decision is that if a sensed value x on a sensing node is *close* to the predicted value $\hat{x}$ (predicted locally on the sensing node), there is not much benefit by delivering x to the node dedicating for further processing, i.e., the EN in our case. Otherwise, it is important for the EN to consider x to proceed with accurate analytics tasks. Evidently, there is a trade-off, that we should pay attention, between contextual data communication and accuracy of analytics due to approximation. On the one hand, by selectively sending and receiving contextual data within an edge network increases the network life time and the available bandwidth, since less data are circulated. On the other hand, this comes at the expense of quality of the analytics tasks, due to local processing with *deliberately* approximated data at the EN. This requires that:

- The EN employs a mechanism to re-construct the undelivered data in light of proceeding with processing and analytics tasks.
- The SAN is equipped with computation capacity for real-time prediction of $\hat{x}$. This requirement is already provided by EC environments, where IoT devices are armed with both: sensing and computing capabilities.

In this paper we propose an intelligent mechanism that makes use of *all* available resources and capabilities in the edge network to support edge analytics.

2.3 Contribution and Organization

The principle of selective data delivery for supporting edge analytics is not directly adopted. Though, it should be adjusted to *split* the predictive intelligence to the SAN and the EN: the former node *locally predicts* the expected data and *locally decides* on their delivery given an approximation error bound; the latter node *locally predicts/reconstructs* the undelivered data in case that the SAN decided not to deliver. Through this intelligence split, we introduce a mechanism to support all the above mentioned edge-analytics methodologies. That is, such predictive intelligence is applied before a SAN needs to communicate with an EN for an analytics task, while the EN should re-construct the un-delivered data before proceeding with the scheduled analytics task. Should the IoT applications tolerate some *error* in the derived analytics task, like prediction accuracy and quality of aggregation and data fusion operators, then this mechanism can be communication efficient hired for edge analytics, as will be shown in our performance Sect. 5. We show that our mechanism can provide aggregation analytics tasks in the edge network. The fundamental question that we pose in this work is: *Is predictive intelligence on SANs and ENs efficient by trading quality of aggregation analytics for communication saving?* In this paper **we study the impact of this mechanism on edge analytics** in terms of efficiency by sacrificing quality of analytics results. The aim is to achieve a significant increase in the edge network lifetime by tolerating quality of the analytics tasks.

To the best of our knowledge this is a first mechanism that explores the potentials of predictive intelligence on the EC paradigm under the concept of analytics to the edge. The key contributions in this paper are:

- We present a decentralized predictive intelligence mechanism within an edge network with SANs and ENs for supporting edge analytics in a communication efficient way;
- The proposed mechanism is evaluated for showcasing the trade-off between accuracy (quality) of edge aggregation analytics operators and communication overhead;
- We propose certain light-weight re-construction policies for ENs with $O(d)$ computational complexity in a d-dimensional data space using exponential smoothing as a selective data delivery mechanism.
- We evaluate the mechanism using real contextual data from sensors and actuators networks.

The paper is organized as follows: In Sect. 3 we present our rationale and basic concept of the edge predictive intelligence formulated by certain definitions, preliminaries, and the fundamental metrics for evaluating the proposed mechanism. Section 4 reports on the predictive intelligence split to the SAN and EN perspectives elaborating on certain policies for data delivery and re-construction. In Sect. 5 we showcase the performance of the proposed mechanism with a real contextual dataset, while Sect. 6 concludes the paper with future research agenda on edge analytics.

3 Edge Predictive Intelligence

3.1 Rationale

We consider an edge network with connected ENs forming an arbitrary topology. Each EN j is connected with n_j SANs in a tree-like topology with root the EN and leaves its SANs as shown in Fig. 1. A SAN i is connected with its unique EN j and $\mathcal{N}_j = \{1, \dots, n_j\}$ denotes the SAN set of the EN j, i.e., $i \in \mathcal{N}_j$.

A SAN i at every time instance $t = 1, 2, \dots$ senses a d-dimensional row vector $\mathbf{x}_t = [x_{1t}, \dots, x_{dt}] \in \mathbb{R}^d$ of contextual parameters, like temperature, humidity, sound, wind speed, air pollutant chemical compounds, etc. Hereinafter, we refer to $\mathbf{x}$ as *context vector*. The SAN i can communicate with its EN j in the edge network by

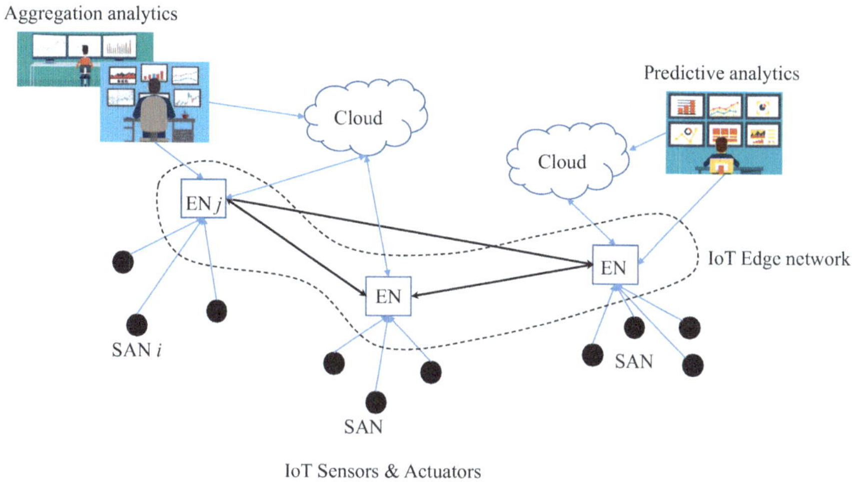

Fig. 1 The edge network with the ENs and the corresponding SANs provide communication efficient analytics to end-users, analysts, and to IoT applications

transferring context vectors. To materialize the proposed predictive intelligence, the SAN i is equipped with a context vector prediction algorithm $f_i(\mathbf{x}_{t-1}, \ldots, \mathbf{x}_{t-N})$, which uses the recent past $N \geq 1$ sensed context vectors stored in a sliding window $\mathcal{W}$ of size N to predict the context vector $\hat{\mathbf{x}}_t$ at time instance t. That is:

$$\hat{\mathbf{x}}_t = f_i(\mathbf{x}_{t-1}, \ldots, \mathbf{x}_{t-N}) = f_i(\mathcal{W}) \tag{1}$$

and window $\mathcal{W} = (\mathbf{x}_{t-N}, \ldots, \mathbf{x}_{t-1})$. The SAN i, after actually sensing the context vector $\mathbf{x}_t$ at time t, locally predicts the predicted context vector $\hat{\mathbf{x}}_t$, thus, the **local prediction error** is:

$$e_t = \|\mathbf{x}_t - \hat{\mathbf{x}}_t\| \tag{2}$$

where $\|\mathbf{x}\| = (\sum_{k=1}^{d} x_k^2)^{1/2}$ is the Euclidean norm of $\mathbf{x}$. Such prediction capability yields the SAN able to decide whether to send context vectors $\mathbf{x}$ to its EN j or not for further processing. SAN i relies on a θ-based context vector delivery decision rule:

- **Case 1** If the predicted $\hat{\mathbf{x}}_t$ differs from the actual sensed $\mathbf{x}_t$ with respect to a *decision threshold* $\theta > 0$, i.e., $e_t > \theta$, then the SAN i sends the actual $\mathbf{x}_t$ to the EN j.
- **Case 2** Otherwise, i.e., $e_t \leq \theta$, the SAN i does not send $\mathbf{x}_t$ to the EN j. In this case, the EN j is responsible for *reconstructing* a context vector locally for further processing.

In Case 1, the EN j receives the transmitted context vector $\mathbf{x}_t$ from SAN i. In Case 2, the EN j is equipped with a re-construction function

$$\bar{\mathbf{x}}_t = g_j(\mathbf{u}_{t-1}, \ldots, \mathbf{u}_{t-M}) = g_j(\mathcal{W}) \tag{3}$$

of the recent $M \geq 1$ context vectors $\mathbf{u}$ from a sliding window $\mathcal{W} = (\mathbf{u}_{t-M}, \ldots, \mathbf{u}_{t-1})$ to locally predict (reconstruct) the undelivered vector $\mathbf{x}_t$, notated by $\bar{\mathbf{x}}_t$ through historical context vectors. Specifically, the context vectors $\mathbf{u}$ in the EN's sliding window $\mathcal{W}_j$ correspond to either the actual received context vectors $\mathbf{x}$ from the SAN i (Case 1) or the past locally re-constructed context vectors $\bar{\mathbf{x}}$ from g_j (Case 2), i.e.,

$$\mathbf{u}_t = \begin{cases} \mathbf{x}_t & \text{if } e_t > \theta \text{ (Case 1)} \\ \bar{\mathbf{x}}_t = g_j(\mathcal{W}) & \text{, otherwise; (Case 2)} \end{cases}$$

The **re-construction difference** at the EN j is then:

$$a_t = \begin{cases} 0 & \text{Case 1,} \\ \|\mathbf{x}_t - \bar{\mathbf{x}}_t\| & \text{Case 2.} \end{cases} \tag{4}$$

The sliding window at SAN i contains only actual (sensed) context vectors $\mathbf{x}$, while the sliding window at EN j contains either actual context vectors $\mathbf{x}$ (received from SAN i) or re-constructed context vectors $\bar{\mathbf{x}}$ locally generated by EN j. The difference (norm) between the predicted context vector $\hat{\mathbf{x}}$ on SAN i and the reconstructed context vector $\bar{\mathbf{x}}$ at EN j is $\|\hat{\mathbf{x}} - \bar{\mathbf{x}}\| = \|\mathbf{e} - \mathbf{a}\|$, with $\mathbf{a} = \bar{\mathbf{x}} - \mathbf{x}$ and $\mathbf{e} = \hat{\mathbf{x}} - \mathbf{x}$. This difference is zero when both the predictor and the re-constructor on SAN i and EN j, respectively, result in the same *error*. Overall, when $e_t > \theta$, the reconstruction difference $a_t = 0$, while when $e_t \leq \theta$, the reconstruction difference $a_t \geq 0$.

Goal: Given a decision threshold $\theta > 0$ at SAN i, we study the performance of certain aggregation analytics tasks on EN j. We qualitatively derive sufficient conditions for this and reveal that the decision is a function of both the desired error bound and the correlation among the sensed contextual data values. When the decision threshold is very tight or the correlation is not significant, the SAN i always has to send its context vectors to the EN j. Due to the characteristics and inherent dynamics of the SANs' contextual data, when the underlying data distribution evolves over time, prediction techniques may not work efficiently for a set of less predictable contextual data. We provide certain definitions and preliminaries before elaborating on our distributed intelligence mechanism.

3.2 Definitions and Problem Formulation

Definition 1 (*Sliding Window*) A sliding window $\mathcal{W}$ is specified by a fixed-size temporal extent $N > 0$ ('horizon') by appending new context vectors and discarding older ones on the basis of their appearance.

For instance, at time t, a sliding window $\mathcal{W}$ is a sequence of all context vectors observed from $t - N$ to $t - 1$, i.e., $\mathcal{W} = (\mathbf{x}_{t-N}, \mathbf{x}_{t-N+1}, \dots, \mathbf{x}_{t-1})$. As an example, an analytics query over $\mathcal{W}$ could be: 'continuously return all context vectors of the past hour, i.e., $N = 60$ min'. The sliding window is the most widely used in continuous aggregation and fusion analytics functions [33–36].

The **aggregation analytics** tasks are evaluated over the contents of a window $\mathcal{W}$. The aggregated results change over time as the window slides. We use the classification from [37] that divides aggregation functions into three categories: distributive, algebraic, and holistic. Let $\mathcal{W}$, $\mathcal{W}_1$, and $\mathcal{W}_2$ be windows. An aggregation analytics function $h : \mathcal{W} \rightarrow \mathbb{R}^d$ is distributive if $h(\mathcal{W}_1 \cup \mathcal{W}_2)$ can be computed from $h(\mathcal{W}_1)$ and $h(\mathcal{W}_1)$ for all $\mathcal{W}_1$, $\mathcal{W}_2$. An aggregation analytics function h is algebraic if there exists a 'synopsis function' σ such that for all $\mathcal{W}$, $\mathcal{W}_1$, and $\mathcal{W}_2$: (1) $h(\mathcal{W})$ can be computed from $\sigma(\mathcal{W})$; (2) $\sigma(\mathcal{W})$ can be stored in constant memory; and (3) $\sigma(\mathcal{W}_1 \cup \mathcal{W}_2)$ can be computed from $\sigma(\mathcal{W}_1)$ and $\sigma(\mathcal{W}_2)$. An aggregation analytics function h is holistic if it is not algebraic. Among the standard aggregates, `MAX` and `MIN` are distributive, `AVG` is algebraic, since it can be computed from a synopsis containing `SUM` and `COUNT`, and `QUANTILE`, `MEDIAN` are holistic.

Example 1 We can define the `AVG` and `MAX` analytics functions: $h^{avg}(\mathcal{W}) = \frac{1}{N} \sum_{k=t-N}^{t} \mathbf{x}_k$ and $h^{max}(\mathcal{W}) = [\max\{x_{1k}\}, \dots, \max\{x_{dk}\}]_{k=t-N}^{t}$, respectively.

In our case, the aggregation analytics function h is running on EN j *for each* sliding window $\mathcal{W}$ containing M received and/or re-constructed context vectors from the SAN $i \in \mathcal{N}_j$ depending on Case 1 and Case 2. Note that such functions are built-in constructs in IoT-application specific continuous analytics queries.

Example 2 The aggregation analytics query 'every minute find the average temperature and the maximum humidity over context streams 'temperature' and 'humidity' collected during the past hour' in Continuous Query Language [38] involving `AVG` and `MAX` operators in a sliding window $\mathcal{W}, N = 60$ min can be expressed as follows:

```
SELECT AVG(temperature), MAX(humidity)
FROM Context Streams [RANGE 60 MINUTES SLIDE 1 MINUTE]
```

Note, typical progressive aggregates like `SUM`, `MIN` and `AVG` requires constant time $O(1)$ per value since there is no need to scan the entire window [39, 40]. More advanced aggregation analytics functions like outliers detection or concept drift detection in a sliding window $\mathcal{W}$ require multiple scanning of the $\mathcal{W}$. Aggregation analytics functions can be also combined on a EN to infer certain events that might trigger decision making.

Example 3 Consider the evaluation of a situational context (localized event stream processing) for the past ten minutes as the activation of the following rule with conjunctive predicates associated with AVG and MAX aggregation analytics functions over 'temperature' and 'wind-speed' sliding windows from two corresponding SANs:

```
EVENT := IF AVG(temperature) ≥90 AND MAX(wind-speed) ∈
[10,20]
WITHIN 10 minutes THEN ACTION is 'warning'
```

Definition 2 (*Aggregation Analytics Difference*) Consider an EN j and its SAN $i \in \mathcal{N}_j$. The aggregation analytics difference β_i between the analytics result on EN j derived from aggregation function h over the window $\mathcal{W}$ in the EN j and the *actual* analytics result derived from h over the window $\mathcal{W}^*$, which contains only the actual context vectors from SAN i to EN j (ground truth) is:

$$\beta_i = \|h(\mathcal{W}) - h(\mathcal{W}^*)\|. \tag{5}$$

The aggregation analytics difference β_i denotes how much the aggregation results over the window $\mathcal{W}$ on EN j with context vectors $\mathbf{u}$ *differ* from the aggregation results over the window $\mathcal{W}^*$ with context vectors $\mathbf{x}$, should SAN i have sent all context vectors to EN j. Obviously, if we encounter only the Case 1, then $\beta_i = 0, \forall i \in \mathcal{N}_j$. Now, since we allow SAN i to decide on sensing context vectors w.r.t. θ and EN j being able to re-construct undelivered context vectors, then $\beta_i \geq 0$. The concept is how much an IoT application tolerates this difference in analytics results in light of communication efficiency in the edge network.

Hence, given a decision threshold $\theta > 0$, our aim is to examine the impact of our predictive intelligence mechanism on (i) the re-construction difference a in (4) and (ii) the aggregation analytics difference β in light of communication efficiency by saving significant network bandwidth.

4 Predictive Intelligence Split

The intelligence of the proposed mechanism is split into two parts: (i) the SAN's intelligence with respect to the local prediction algorithm f_i and (ii) the EN's intelligence with respect to the local re-construction algorithm g_j that supports the analytics tasks introduced in Sect. 3.2.

4.1 Sensor-Actuator Node Intelligent Part

Consider a SAN i and let us elaborate on the first part. Very complex prediction models are not practical in the discussed EC paradigm due to the limited (energy-constrained) computational capacity of the SANs. Fortunately, simple linear predictors are sufficient to capture the temporal correlation of realistic contextual data as shown by previous studies [41–43]. A sliding window-based linear predictor is one of popular approaches to predicting the future based on past N measurements.

In this work, we are seeking to reduce the computational power for prediction and to use a small fraction of the SAN's computing power by adopting a predictive function with low complexity and computational effort. Multivariate exponential smoothing, used for time series forecast, is an ideal predictor adopted in our case, as its computational complexity is $O(d)$ in a d-dimensional space (explained later). A simple exponential smoothing weighs the current sensed context vector $\mathbf{x}_t$ and the historic context vectors [44]. This simple smoothing function is adopted as the prediction function f_i for the θ-based decision making.[1]

At each time t, a smoothed context vector $\mathbf{s}_t$ is calculated by using the current sensed context vector $\mathbf{x}_t$ and the previous smoothed vector $\mathbf{s}_{t-1}$, i.e.,

$$\mathbf{s}_t = \alpha \mathbf{x}_t + (1 - \alpha)\mathbf{s}_{t-1} \tag{6}$$

initializing with $\mathbf{s}_0 = \mathbf{x}_0$. The relationship between the history of the measured data and the current data is represented by $\alpha \in [0, 1]$. A higher α denotes more importance to the current values and less importance to the historic values. Normally, $\alpha = 0.7$ [44]. The calculated smoothed vector $\mathbf{s}_{t-1} = [s_{1,t-1}, \ldots, s_{d,t-1}]$ refers to the predicted context vector $\hat{\mathbf{x}}_t$, that is: $\hat{\mathbf{x}}_t = f_i(\mathcal{W}_i) = \mathbf{s}_{t-1}$, with the window $\mathcal{W} = (\mathbf{s}_{t-1})$ at SAN i containing only the recent smoothed context vector. Hence, the complexity of f_i is $O(d)$; we require d computations for smoothing the $\mathbf{s}_t$ in (6) at time instance t. The forwarding decision of the actual $\mathbf{x}_t$ to the EN j depends whether the prediction error $e_t = \|\mathbf{s}_{t-1} - \mathbf{x}_t\|$ exceeds the threshold θ.

4.2 Edge Node Intelligent Part

On the other side, the EN j, at time instance t either receives $\mathbf{x}_t$ (Case 1) or nothing (Case 2). In Case 1, the EN j simply inserts the delivered $\mathbf{x}_t$ into its corresponding window $\mathcal{W}$ (which is associated with the SAN $i \in \mathcal{N}_j$) discarding the oldest context vector, i.e., $\mathbf{u}_t = \mathbf{x}_t$. In Case 2, the EN j encounters an undelivered vector problem, since there is nothing to push in the sliding window $\mathcal{W}$. Such undelivered context vectors must be re-constructed with the available context vectors $\mathbf{u}$ reside currently in the $\mathcal{W}$ at EN j. In order to achieve this, we propose three re-construction policies,

[1]Double exponential smoothing (Holt-Winters time series smoothing) could be adopted dealing with the same computational complexity.

i.e., variants of the re-construction function $g_j(\mathcal{W})$. We should stress that, we require a computationally efficient re-construction function on EN j, thus, being relatively a small overhead compared to the analytics tasks. Those policies are discussed below.

Policy 1 This policy, in Case 2, uses the *most* recent context vector from $\mathcal{W}$ at EN j, i.e., the first element of the sliding window, as the re-constructed context vector. Therefore, the re-constructed context vector is inserted into the $\mathcal{W}$ and the oldest context vector from the window is discarded. Note that, after this insertion, there are two duplicates of the most recent context vector in the window. There might be also the case where the entire window (of length N) would have contained the same context vector if the SAN i had not sent a context vector in the last N time instances. This denotes that, during this recent history of N time instances, the maximum difference of the sequentially sensed context vectors measured on SAN i is less that θ. In this case, it is redundant to send *similar* context vectors to EN j given a threshold θ. In Case 1, the EN j simply inserts the delivered $\mathbf{x}_t$ into the window and discards the oldest context vector.

Policy 2 This policy, in Case 2, re-constructs the undelivered context vector $\bar{\mathbf{x}}_t$ as the average vector of the current context vectors in the window $\mathcal{W}$, i.e.,

$$\bar{\mathbf{x}}_t = g_j(\mathcal{W}) = \frac{1}{N} \sum_{k=t-N}^{t-1} \mathbf{u}_k.$$

This re-constructed value is then inserted into the window discarding the oldest one. In Case 1, the EN j simply inserts the delivered $\mathbf{x}_t$ into the window and discards the oldest context vector.

Policy 3 This policy applies the exponential smoothing algorithm (discussed above) for re-constructing the undelivered context vector in the EN j. In Case 1, the EN j simply inserts the delivered $\mathbf{x}_t$ into the window and discards the oldest context vector. Moreover, after this insertion, the EN j calculates the smoothing context vector $\mathbf{s}'_t$ based on the delivered $\mathbf{x}_t$ and the previously calculated smoothed context vector, i.e.,

$$\mathbf{s}'_t = \alpha \mathbf{x}_t + (1-\alpha)\mathbf{s}'_{t-1}.$$

In Case 2, this policy re-constructs the $\bar{\mathbf{x}}_t$ with the recently smoothed context vector $\mathbf{s}'_{t-1}$ (exploiting the context vector delivery in Case 1) and discards the oldest context vector from the window. Note that, the series of the smoothed vectors $\mathbf{s}'_t$ in EN j is not the same with the series of the smoothed vectors $\mathbf{s}_t$ in SAN, since the vectors $\mathbf{s}'_{t-1}, \mathbf{s}'_{t-2}, \ldots$ are calculated by the $\mathbf{u}_{t-1}, \mathbf{u}_{t-2}, \ldots$ vectors from the window $\mathcal{W}_i$ on EN j. Moreover, in Case 2, after the re-construction of $\bar{\mathbf{x}}_t$ with $\mathbf{s}'_{t-1}$, the smoothed context vector for time instance t is $\mathbf{s}'_t = \alpha\bar{\mathbf{x}}_t + (1-\alpha)\mathbf{s}'_{t-1} = \mathbf{s}'_{t-1}$. Overall, Policy 3 at the EN j has as follows:

$$\begin{cases} \mathbf{s}'_t = \alpha \mathbf{x}_t + (1-\alpha)\mathbf{s}'_{t-1} & \text{, Case 1} \\ \bar{\mathbf{x}}_t = \mathbf{s}'_{t-1} \text{ and } \mathbf{s}'_t = \mathbf{s}'_{t-1} & \text{, Case 2.} \end{cases} \tag{7}$$

5 Performance Evaluation

5.1 *Dataset and Experimental Setup*

In our experiments, we used a real dataset for assessing the performance of the proposed edge prediction intelligence mechanism. The contextual dataset (DS1) was adopted by UCI [45]. This dataset contains twelve SANs of chemical compounds and environmental parameters: CO, PT08.S1 (tin oxide), Non Metanic HydroCarbons, Benzene, PT08.S2 (titania), NOx, PT08.S3 (tungsten oxide), NO_2, PT08.S4, PT08.S5 (indium oxide), temperature, relative humidity, and absolute humidity. All these contextual parameters are required to measure the air pollution of a specific area. These data are collected every hour and refer to $T = 9357$ multi dimensional measurements ($d = 12$) with $n = 12$ SANs and one EN. Inside this dataset missing values occur. For each SAN, we impute those missing values by adopting the missing value imputation method of linear interpolation. This method exploits two data points (x_0, y_0) and (x_1, y_1) to reconstruct a linear function to find for a specific x value the missing value y as follows: $y = y_0 + \frac{y_1-y_0}{x_1-x_0}(x - x_0)$.

For comparison and reproduction, DS1 is *normalised* and *scaled*, i.e., each contextual parameter $x \in \mathbb{R}$ is mapped to $\frac{x-\mu}{\sigma}$ with mean value μ and variance σ^2, and scaled in [0,1] using $\frac{\max\{x\}-x}{\max\{x\}-\min\{x\}}$.

For sensitivity analysis of our mechanism, the experiments were carried out with different values of the decision threshold $\theta \in \{10^{-5}, 10^{-3}, 10^{-2}, 0.05, 0.06\}$. Using a normalised and scaled dataset, θ is interpreted as the percentage change of a measured/sensed value x by: 0.0002%, 0.02%, 2%, 10% and 12% respectively for the chosen θ values, respectively. The chosen θ values are examined for the impact of the local prediction error e_t in (2) on our mechanism to decide on forwarding a measurement. The local predictor/exponential smoother in the SAN side, in our experiment, uses $\alpha \in \{0.5, 0.7, 1\}$ as suggested in [44]. Moreover, in Policy 3, the reconstruction smoother function in EN side uses $\alpha = 0.7$. The sliding window size is set to $N = 10$. This selected size represents for DS1 a history of the last 10 hours.

Overall, our experimental set up includes five θ values and three α values over three different policies (Policy 1, 2, and 3) for reconstruction on the EN. This leads to an overall of 75 experiments for each of the aggregation analytics function $h(\mathcal{W})$: i.e., `AVG`, `MAX` and `MIN` analytics function for evaluation. In order to objectively assess the performance of our mechanism, we implemented the baseline mechanism. This mechanism is produced by capturing the continuous contextual data and transmitting them from all SANs to an EN, without any predictive intelligence on SAN or EN.

5.2 Performance Metrics

We firstly define the performance metrics of **counter of communications**, i.e., the number of sensed values are sent from a SAN i to its EN j. By adopting the baseline mechanism the number of communications is allocated with $12 \cdot 9357 = 112,284$. To better illustrate and compare our mechanism with the baseline, the counter of communications is indicated as 100%. Using our mechanism, the counter is only increasing if θ is exceed and values are transmitted from the SANs to the EN. The overall communication of n SANs over T sensing values, in our method is then:

$$c(T) = \sum_{t=1}^{T} \sum_{i=1}^{n} I_{i,t}, \tag{8}$$

with $I_{i,t} = 1$ if SAN i sends its sensed value to the EN; otherwise $I_{i,t} = 0$. Evidently, the overall communication of n SANs over T sensing values, in the baseline method is $T \cdot n$, since $I_{i,t} = 1, \forall i, t$. The percentage of communication is then $\frac{c(T)}{Tn}$.

The re-construction difference a in (4), and the aggregation analytics difference β in (5) is evaluated using the **symmetric mean absolute percentage error (SMAPE)**. During the experiments, we calculated the average SMAPE per SAN for each time instance $t \in \{1, \ldots, T\}$. This metric is used to represent a percentage error of SMAPE $\in [0, 100]$ because of its unbiased properties [46]. SMAPE is defined for reconstruction difference a and aggregation analytics difference β as:

$$\text{SMAPE} = \begin{cases} \frac{100}{T} \sum_{t=1}^{T} \frac{a_t}{\|\mathbf{x}_t\| + \|\bar{\mathbf{x}}_t\|} & , \|\mathbf{x}_t\| + \|\bar{\mathbf{x}}_t\| > 0 \text{ for } a, \\ \frac{100}{T} \sum_{t=1}^{T} \frac{\beta_t}{\|h(\mathcal{W})\| + \|h(\mathcal{W}^*)\|} & , \|h(\mathcal{W})\| + \|h(\mathcal{W}^*)\| > 0 \text{ for } \beta. \end{cases} \tag{9}$$

Our major aim is to demonstrate the trade-off between communication and re-construction difference/aggregation analytics difference. That is, we can tolerate some *slight* increase in the analytics error by gaining a *significant decrease* in the number of communications, thus, being communication efficient and prolonging the edge network lifetime. It is evaluated how the aggregated analytics functions $h(\mathcal{W})$ and re-construction inside an EN behave with decreasing the number of communications towards the EN. This decrease of communications is produced by increasing the value of θ and changing the value of α inside the SAN (exponential smoother).

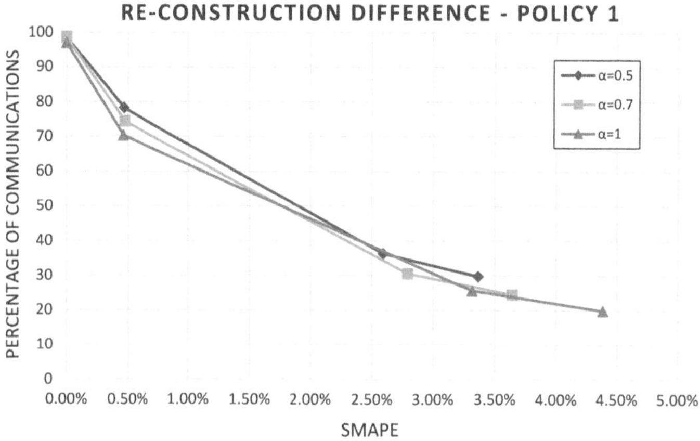

Fig. 2 Re-construction Difference trade-off for Policy 1; percentage of communication against SMAPE

5.3 *Performance Assessment*

Performing our method of predictive intelligence with the explained chosen values, we can illustrate the trade-off between communication and error for the re-construction and aggregation analytics differences. Independent of the specific differences, our aim is to reduce the percentage of communication only with a slight increase of the error.

Generally, increasing the value of θ is decreasing the number of communications towards the EN. The reason behind this is that θ is demonstrating the tolerance for a change of the expected value and the actual/sensed one. Therefore high values of θ are indicating that values can vary between a larger range before they are sent towards the EN. Furthermore, it is notable that the number of communications is highly dependent on the exponential smoother parameter α. Given the same θ value, the number of communications is decreasing with higher values for α. Increasing α means reducing the influence of previous/historical measured data and weight the current data high. Having a value for $\alpha = 1$, the current measured value is compared against only the previous for a forwarding decision.

5.3.1 Re-Construction Difference Assessment

Evaluating the re-construction difference for DS1 it is applicable that the chosen reconstruction policy inside the EN is highly influencing the produced SMAPE. Given the three different policies, Policy 2 is generating the highest error values over $\alpha \in \{0.5, 0.7, 1\}$ and $\theta \in \{10^{-5}, 10^{-3}, 10^{-2}, 0.05, 0.06\}$. Policy 3 as reconstruction method is creating nearly the same error percentage as Policy 1. Policy 1 represents the best possible solution for reconstruction the undelivered values inside the EN.

Comparing Policy 1 against different values for α it can be observed from Fig. 2, that increasing α is increasing the SMAPE. Considering the trade-off between SMAPE and percentage of communication, higher values for α illustrate a lower SMAPE with less communication. Figure 2 shows the re-construction difference for Policy 1 over different α values. His worth noting that, as shown in Fig. 2, using our proposed predictive intelligence for edge analytics on DS1, a communication overhead of 30% can be saved by tolerating an re-construction error of less than 1%. If IoT applications can tolerate up to 2% error in their analytics accuracy, it is possible to save between 50 and 60% of communication. Saving this amount of communication with a given tolerated error would increase the lifetime of an edge network between 30 and 50%.

5.3.2 Aggregation Analytics Assessment

Besides the re-construction difference, the aggregation analytics difference produced by our proposed method is an important metric by many analytics IoT applications. From our experiments it is applicable for all three aggregation functions, `AVG`, `MIN` and `MAX`, that Policy 2 is producing the highest SMAPE for the aggregation analytics difference. Similar to the re-construction difference, Policy 1 is generating the lowest average error per SAN over the entire time frame T. In our experiments a value of $\alpha = 0.5$ is producing the lowest error over all three aggregation functions by employing the re-construction Policy 3. Comparing Policy 1 over all three aggregation functions, Fig. 3 shows this for DS1. Specifically, one can observe from Fig. 3 that, similar to the re-construction difference, the aggregation analytics difference depends on α. Higher values of α producing a better trade-off. For `MIN` and `MAX` this reverses after a SMAPE of around 1.5%. `AVG` continuously produces a good trade-off for high α.

For DS1 it is observed in Fig. 3, that a reduction of 20% by communication for `MIN`, `MAX` and 30% for `AVG` is only generating an error of 0.5%. Therefore, it is possible to increase the lifetime of a network up to 30% with tolerating a slightly difference towards the true result. For applications that could tolerate a higher discrepancy for this kind of aggregation functions, they can save up to 50% with an error of 1.5–2%.

6 Conclusions

We focus on the edge computing paradigm where pushing aggregation analytics to the edge of the IoT network allows the complexity of analytics tasks to be distributed into many smaller and more manageable pieces and to be physically located at the source of the contextual information it needs to work on. We introduce a lightweight, distributed, predictive intelligence mechanism that supports communication efficient aggregation analytics within the edge network of SANs and ENs. The mechanism

Fig. 3 Aggregation Analytics Difference trade-off for DS1 with Policy 1; percentage of communication against SMAPE

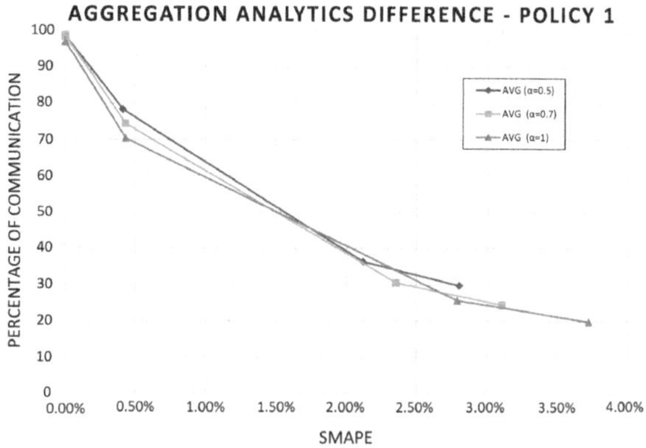

(a) AVG function

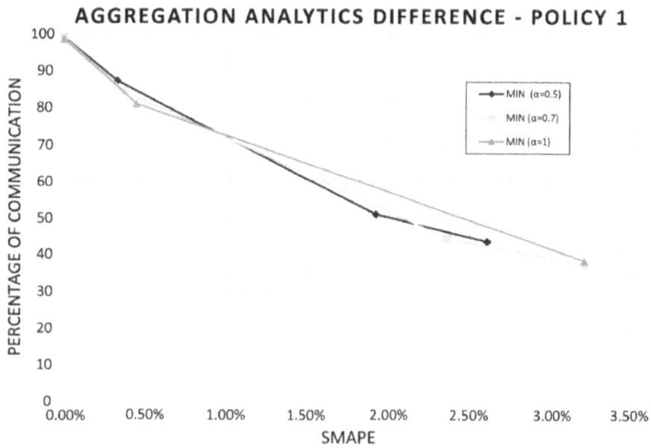

(b) MIN function

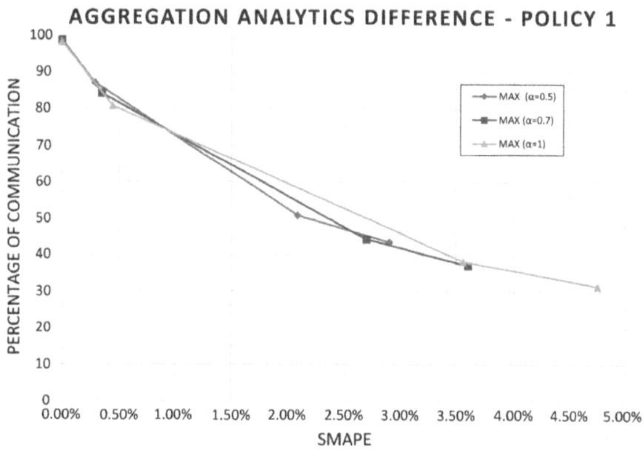

(c) MAX function

is based on the idea of locally deciding whether to deliver contextual data or not in light of minimizing the induced communication overhead and providing high quality analytics tasks. Based on splitting this intelligence into: prediction (through exponential smoothing) and decision making at the SANs and context re-construction at ENs (by proposing three policies), we eliminate data transfer at the edge of the network, by exploiting the predictability of the captured contextual data. We provide comprehensive experimental evaluation of the proposed mechanism over a real multidimensional contextual dataset for aggregation analytics tasks and show the benefits stemmed from its adoption in edge computing environments. We experimented with the trade-off between accuracy (quality) of edge analytics tasks and communication overhead. Our mechanism demonstrated its efficiency in supporting high quality of edge analytics by tolerating a relatively low error in light of decreasing significantly the communication overhead in an edge network. Our future agenda includes certain modifications of our mechanism to support predictive analytics tasks including outliers detection and linear regression predictive models in edge computing environments.

References

1. M. Satyanarayanan et al., Edge Analytics in the Internet of Things, in *IEEE Pervasive Computing*, vol. 14, no. 2. pp. 24–31, Apr-June 2015
2. The mobile-edge computing initiative, http://www.etsi.org/technologies-clusters/technologies/mobile-edge-computing
3. I. Stojmenovic, S. Wen, The Fog computing paradigm: scenarios and security issues, in *2014 Federated Conference on Computer Science and Information Systems, Warsaw*, 2014, pp. 1–8
4. S. Yi, C. Li, Q. Li, A survey of fog computing: concepts, applications and issues, in *Proceedings of the 2015 Workshop on Mobile Big Data*, 2015, pp. 37–42
5. A. Vulimiri, C. Curino, P.B. Godfrey, T. Jungblut et al., WANalytics: geo-distributed analytics for a data intensive world, in *Proceedings of the 2015 ACM SIGMOD International Conference on Management of Data*, 2015, pp. 1087–1092
6. B. Cheng, A. Papageorgiou, M. Bauer, Geelytics: enabling on-demand edge analytics over scoped data sources, in *IEEE International Congress on Big Data (BigData Congress)*. San Francisco, CA, vol. 2016, pp. 101–108, 2016
7. R. Ganti, F. Ye, H. Lei, Mobile crowdsensing: current state and future challenges. Commun. Mag. IEEE **49**(11), 32–39 (2011)
8. N. Lane, E. Miluzzo, H. Lu, D. Peebles, T. Choudhury, A. Campbell, A survey of mobile phone sensing. Commun. Mag. IEEE **48**(9), 140–150 (2010)
9. L.M. Oliveira, J.J. Rodrigues, Wireless sensor networks: a survey on environmental monitoring. J. Commun. **6**(2), 143–151 (2011)
10. J. Kang et al. High-fidelity environmental monitoring using wireless sensor networks, *Proceedings of the 11th ACM Conference on Embedded Networked Sensor Systems (SenSys '13)* (USA, Article 67, 2013)
11. A. Awang et al. RIMBAMON: a forest monitoring system using wireless sensor networks, in *ICIAS 2007*, pp. 1101–1106, 2007
12. E. Zervas et al., Multisensor data fusion for fire detection. Inf Fusion Elsevier **12**(3), 1566–2535 (2011)
13. C. Anagnostopoulos, S. Hadjiefthymiades, K. Kolomvatsos, Accurate, dynamic, and distributed localization of phenomena for mobile sensor networks. ACM Trans. Sen. Netw. **12**, 2, Article 9 (April 2016), 59 pages (2016)

14. S. Nittel, A Survey of geosensor networks: advances in dynamic environmental monitoring. Sensors **9**, 5664–5678 (2009)
15. Yu. Pieter Simoens, P. Pillai Xiao, Z. Chen, K. Ha, M. Satyanarayanan, Scalable crowd-sourcing of video from mobile devices, in *Proceeding of the 11th Annual International Conference on Mobile Systems, Applications, and Services (MobiSys '13)* (ACM, New York, NY, USA, 2013), pp. 139–152
16. G. Xu et al., Applications of wireless sensor networks in marine environment monitoring: a survey. Sensors **14**(9), 16932–16954 (2014)
17. G.W. Eidson et al., The south carolina digital Watershed: end-to-end support for realtime management of water resources, in *Proceedings of the 4th International Symposium on Innovations and Real-time Applications of Distributed Sensor Networks (IRADSN 09)*, vol. 2010 (USA, May 2009)
18. N. Nguyen et al. A real-time control using wireless sensor network for intelligent energy management system in buildings, in *Proceedings of the IEEE Worskshop on Environmental Energy and Structural Monitoring Systems (EESMS 10)*, pp. 87–92, Sept 2010
19. K. Kolomvatsos; C. Anagnostopoulos; S. Hadjiefthymiades, Data fusion and type-2 fuzzy inference in contextual data stream monitoring, in *IEEE Transactions on Systems, Man, and Cybernetics: Systems*, vol. PP, no. 99, pp. 1–15, June 2016
20. S.M. McConnell, D.B. Skillicorn, A distributed approach for prediction in sensor networks, in *Proceedings of the SIAM International Conference on Data Mining Workshop Sensor Networks*, 2005
21. D. Tulone, S. Madden, An energy-efficient querying framework in sensor networks for detecting node similarities, in *Proceedings of the IEEE/ACM International Conference on Modeling, Analysis, and Simulation of Wireless and Mobile Systems (MSWiM)*, 2006
22. S. Goel, T. Imielinski, Prediction-based monitoring in sensor networks: taking lessons from MPEG. ACM SIGCOMM Comput. Commun. Rev. **31**(5), 82–98 (2001)
23. A. Simonetto, G. Leus, Distributed maximum likelihood sensor network localization, in *IEEE Transactions on Signal Processing*, vol. 62, no. 6, pp. 1424–1437, 15 Mar 2014
24. G. Kejela, R.M. Esteves, C. Rong, *Predictive Analytics of Sensor Data Using Distributed Machine Learning Techniques*, pp. 626–631, 2014
25. R. Gemulla, E. Nijkamp, P.J. Haas, Y. Sismanis, Large-scale matrix factorization with distributed stochastic gradient descent, in *Proceedings of the 17th ACM SIGKDD International Conference on Knowledge Discovery and Data Mining (KDD '11)* (ACM, New York, NY, USA, 2011), pp. 69–77
26. C. Anagnostopoulos, S. Hadjiefthymiades, Advanced principal component-based compression schemes for wireless sensor networks. ACM Trans. Sen. Netw. **11**, 1, Article 7, 34 pages (2014)
27. C. Anagnostopoulos, S. Hadjiefthymiades, A. Katsikis, I. Maglogiannis, Autoregressive energy-efficient context forwarding in wireless sensor networks for pervasive healthcare systems. Pers. Ubiquitous Comput. **18**(1), 101–114 (2014)
28. C. Anagnostopoulos, S. Hadjiefthymiades, P. Georgas, PC3: principal component-based context compression. J. Parallel Distrib. Comput. **72**(2), 155–170 (2012)
29. A. Manjeshwar, D.P. Agrawal, TEEN: a routing protocol for enhanced efficiency in wireless sensor networks, in *Proceedings of the 15th International Parallel & Distributed Processing Symposium (IPDPS '01)* (IEEE Computer Society, Washington, DC, USA), p. 189
30. K. Papithasri, M. Babu, Efficient multihop dual data upload clustering based mobile data collection in Wireless Sensor Network, in *2016 3rd International Conference on Advanced Computing and Communication Systems (ICACCS), Coimbatore*, pp. 1–6, 2016
31. H. Jiang, S. Jin, C. Wang, Prediction or not? an energy-efficient framework for clustering-based data collection in wireless sensor networks. IEEE Trans. Parallel Distrib. Syst. **22**(6), 1064–1071 (2011). June
32. L. Bottou, F.E. Curtis, J. Nocedal, Optimization Methods for Large-Scale Machine Learning, arXiv:1606.04838
33. M. Dallachiesa, G. Jacques-Silva, B. Gedik, K.-L. Wu, T. Palpanas, Sliding windows over uncertain data streams. Knowl. Inf. Syst. (2014)

34. K. Patroumpas, T. Sellis, Maintaining consistent results of continuous queries under diverse window specifications. Inf. Syst. **36**(1), 42–61 (2011)
35. D.J. Abadi, D. Carney, U. Cetintemel, M. Cherniack, C. Convey, S. Lee, M. Stonebraker, N. Tatbul, S. Zdonik, Aurora: a new model and architecture for data stream management. VLDB J. **12**(2), 120–139 (2003)
36. D.J. Abadi, Y. Ahmad, M. Balazinska, U. Cetintemel, M. Cherniack, J.-H. Hwang, W. Lindner, A.S. Maskey, A. Rasin, E. Ryvkina, N. Tatbul, Y. Xing, S. Zdonik, The design of the Borealis stream processing engine, in *CIDR*, Jan 2005
37. J. Gray, S. Chaudhuri et al., Data cube: a relational aggregation operator generalizing group-by, cross-tab, and sub totals. Data Min. Knowl. Discov. **1**(1), 29–53 (1997). Mar
38. A. Arasu, S. Babu, J. Widom, The CQL continuous query language: semantic foundations and query execution. VLDB J. **15**(2), 121–142 (2006). June
39. K. Patroumpas, T. Sellis, Multi-granular time-based sliding windows over data streams, in *2010 17th International Symposium on Temporal Representation and Reasoning (TIME)*, pp. 146–53, 2010
40. K. Patroumpas, T. Sellis, Window specification over data streams, in *Proceedings of the International Conference on Current Trends in Database Technology (EDBT'06)* (Springer, Berlin, 2006), pp. 445–464
41. D. Chu, A. Deshpande, J.M. Hellerstein, W. Hong, Approximate data collection in sensor networks using probabilistic models, in *Proceedings of the IEEE International Conference on Data Engineering (ICDE)*, 2006
42. V.P. Chowdappa, C. Botella, B. Beferull-Lozano, Distributed clustering algorithm for spatial field reconstruction in wireless sensor networks, in *IEEE 81st vehicular technology conference (VTC Spring), Glasgow*, vol. 2015, pp. 1–6, 2015
43. C. Anagnostopoulos, T. Anagnostopoulos, S. Hadjiefthymiades, An adaptive data forwarding scheme for energy efficiency in Wireless Sensor Networks, in *5th IEEE International Conference Intelligent Systems* (London, 2010), pp. 396–401
44. J. Durbin, S. Jan Koopman, *Time Series Analysis by State Space Methods* (Oxford Statistical Science Series, 2012)
45. S. De Vito, E. Massera, M. Piga, L. Martinotto, G. Di Francia, On field calibration of an electronic nose for benzene estimation in an urban pollution monitoring scenario. Sens Actuators B: Chem. **129**(2), 750–757, 22 Feb 2008. ISSN 0925-4005
46. C. Tofallis, A better measure of relative prediction accuracy for model selection and model estimation. J. Oper. Res. Soc. **66**(8), 1352–1362

Publish-Subscribe Based Monitoring Model for Wireless Sensor Networks

Kemal Cagri Serdaroglu, Tevfik Kadioglu and Sebnem Baydere

Abstract The era of computing based on physical data is coming. IoT gateways are gaining more importance as Internet will be the main infrastructure to gather real world data to augment the pervasiveness in remote monitoring of the physical world. Considering the explosion in the number of connected "things", interconnection models impose more functionality in gateways which traditionally were considered as simple protocol converters. Under this perception, IoT gateways should also operate as platforms for applications that process data, manage data traffic and become an intelligent part of a device-enabled system. In this paper, we give a brief survey of state-of-the interconnection solutions and propose a publish subscribe based service for periodic data acquisition from Wireless Sensor Networks (WSN). The service is designed as an interconnection model over the web server based IoT gateway architecture; Wisegate. The architecture and its services are implemented with a 6LoWPAN border router (6LBR) at the backend and a lightweight stack on sensor nodes (6LN) in order to show its end-to-end operability. The experimental results reveal that the proposed approach has encouraging latency and response time characteristics in the given setup.

Keywords IoT · WSN · TCP/IP · Gateway · 6lbr · 6LoWPAN · Contiki · Low-power · Publish-subscribe · Websocket

K.C. Serdaroglu · S. Baydere (✉)
Department of Computer Engineering, Yeditepe University, 34755 Istanbul, Turkey
e-mail: sbaydere@cse.yeditepe.edu.tr

K.C. Serdaroglu
e-mail: kserdaroglu@cse.yeditepe.edu.tr

T. Kadioglu
Nesta Electronics, 34220 Istanbul, Turkey
e-mail: tk@nestaelektronik.com

R.R. Yager and J. Pascual Espada (eds.), *New Advances in the Internet of Things*, Studies in Computational Intelligence 715, DOI 10.1007/978-3-319-58190-3_3

1 Introduction

Internet of real world data requires highly deployed WSNs which are able to operate with IP networks. The integration of WSN and IoT Platforms open new opportunities for novel applications that meet the needs of modern societies. In this paper, we address WSN interoperability issue and propose a novel publish-subscribe based interconnection solution named as Wisegate-P/S for monitoring wireless sensor networks. The model is an enhancement on the Wisegate architecture presented in [1]. The proposed P/S model provides many-to-many connections between Internet clients and sensor nodes for periodical sensor data acquisition. We have implemented the components of the architecture in an experimental setup utilizing a lightweight protocol stack on sensor nodes to show its advantages over currently available solutions. The rest of the paper is structured as follows; Sects. 2 and 3 discusses common interconnection models and tackles the state-of-the-art solutions to realize these models. Section 4 describes the general structure of the Wisegate architecture and the modules of the proposed interconnection model. Section 5 explains the experimental setup. Section 6 gives a discussion of the results obtained in experiments and concludes the paper with future research directions.

2 Interconnection Models for WSN

In general, two basic approaches and their derivatives realize WSN interoperability with the Internet [2]. The first one is the proxy approach which uses a proxy for interconnection of two networks. While this proxy is the front-end server over Internet for clients, in the back-end, it uses dedicated protocols for WSN to gather physical data. The second one is the sensor node stack based approach which uses end-to-end interconnection scheme between clients and sensor nodes and needs an adaptation gateway and a network stack for conventional protocols required by Internet on a sensor node. In this approach, a WSN can be considered as an internet network and partially, some networking features such as routing, neighbour discovery, network node addressing, network management and some transport layer features such as end-to-end reliable data transfer are added to those communication stacks in WSN nodes.

The approaches described above have advantages and disadvantages. For former, with a proxy as the data server, the sensor data acquisition mechanism does not require IP stack on resource constraint sensor node. Hence, sensor node does not dedicate its computational sources for IP stack activities and WSN only operates the protocols tailored to the constraints of itself. However, this approach lacks a standardized inter-proxy-operation model and proxies require significant semantic translations. A more standardized solution is used by the latter. In this model the common standards offer data and network adaptation over both gateway and sensor nodes. Using these standards, end-to-end interaction between Internet and WSN nodes can easily be achieved. However, IP communication stack requires more computational and memory resources on sensor nodes. This means more power consumption on the WSN side.

The proxy and gateway approaches are implemented with several interconnection models. These models have specific design considerations such as data collection scheme, interconnection method, implementation area of the models, data adaptation and the implementation method in those models. Most relevant state-of-the-art solutions are surveyed in Sect. 3.

3 State-of-the-Art Solutions

Ting et al. [3] use an agent based proxy approach for the interconnection. In this work, user agents are utilized to respond client requests of sensor data. System creates a new user agent for gathering sensor data from WSN. Chen et al. [4] present a prototype of a smart gateway which operates as a proxy for the WSN. The proxy model is divided into two models: simple model and intelligent model. In their simple model, the proxy acts as a forward-server and the proxy only forwards the physical data to the clients. In the second model, proxy has a decision making mechanism and acts as a front-end server. This decision making mechanism is used to make some real time decisions and these decisions are sent to the clients. Santos et al. [5] study a proxy for global IoT architecture for health monitoring and control. This study uses a front-end server and a database for recording and delivering the physical data to various clients such as doctors and pharmacists.

Sensor stack approach is used in several studies for the interoperability. Harvan et al. [6] introduce a 6LoWPAN stack for interconnection specifically designed for TinyOS 2.0 operating system [7] and tested on TelosB and MicaZ sensor nodes. This stack supports UDP transport protocol on sensor nodes and necessary header compression schemes of 6LoWPAN [8]. For the design of the gateway, the adaptation mechanism requires a tunneling daemon and Tun device to bypass link layer frames. Dunkels et al. [9] introduce uIPv6 (microIPv6) and lwIP (lightweight IP) stacks for IP based sensor nodes. Their work is among the first to be recognized for integrating TCP to the sensor nodes for end-to-end reliable communication. When uIPv6 is designed for small devices, lwIP supports extra IPv6 operations such as ICMPv6, ND and DAD. These stacks were integrated in Contiki OS [15].

Afformentioned studies pave the way for some data protocol standards in some recent IoT applications as given in Table 1. Various implementations of these protocols are provided in different software platforms. CoAP (Constrained Application Protocol) [10] is a REST standard for resource constraint devices. It enables end devices to communicate with IP nodes using HTTP messaging. CoAP uses a proxy between constrained environments (e.g. a WSN) and Internet. With the help of the proxy, HTTP server in the Internet cloud can communicate with a resource constraint node. A CoAP proxy operates dual stack for interconnection. It uses UDP as transmission protocol between the gateway and the sensor node. Since the data

Table 1 Comparison of commonly used IoT data protocols

Protocol	CoAP	RESTful HTTP	MQTT	Websocket
Transport layer	UDP	TCP	TCP	TCP
Messaging	Request/response periodic	Request/response	Pub/sub	Asyncronous
WSN compatibility	Very Good	Suitable	Suitable	Suitable
Hardware resource	10 Ks	10 Ks	10 Ks	10 Ks
Community	IPSO, OMA, IETF, nem2m	IETF, oneM2M	IBM	IETF
Security	Opt.DTLS	Opt.SSL/TLS	Opt.SSL/TLS	Opt.SSL/TLS
Latency	Medium	Medium	Good	Good
Cloud compatibility	Good	Good	Good	Medium
Power compatibility	Good	Good	Medium	Good
Device discovery	Very good	Application	Good	Application

transmission with UDP is unreliable, CoAP additionally provides a timeout mechanism for retransmission of lost packets.

As can be seen from the table, CoAP is more efficient than others because it is a UDP-based protocol. However, each node requires a direct connection request from the corresponding user. The power consumption of the node is increased due to the request/response or optional periodic based connection architecture. HTTP also has a similar connection architecture to CoAP, hence has the same disadvantages.

MQTT protocol [11] provides device data collection service for telemetry applications. Its goal is to collect data from many devices and transport data to the IT infrastructure. It targets large networks of small devices that need to be monitored or controlled from the cloud. MQTT has support for publish/subscribe model. However, since the MQTT topic must be placed in each message sent to the server it has an additional overhead which limits its scalability and real time performance [16].

Websocket protocol [12] supports client/server architecture and streaming capability. This provides a basic connection scheme, such as UDP or TCP sockets, as well as the ability to manage all client connections from a single port. It also supports many-to-many connections with low protocol overhead. When using a WSN protocol that does not support IP; such as zigbee, z-wave, or other no-IP WSNs, the websocket connection structure provides convenience in architectural design since multiple connection information can be transmitted through a single interface stream. Hence, our proposed Wisegate P/S architecture uses websocket to access sensor nodes.

4 Wisegate Model

Wisegate has been designed as a service scheme for end-to-end reliable interconnection between clients over the Internet and sensor nodes. It uses a web server to handle the request/response mechanism of multiple clients and maintains the end-to-end connections between the server and clients. In addition, it utilizes a packet adaptation layer which transforms the request messages from a client to a byte coded representation and decode the byte-coded message from sensors to generate a response for the client. For the adaptation of sensor nodes to the gateway, a lightweight middleware is designed on sensor nodes. This middleware decodes the byte-coded requests from an IP client and generates a byte-coded response to the gateway. The details of the conversion protocol is given in [1].

In Wisegate, the service scheme described above is implemented with two interacting modules. The former is the gateway application in the front-end of the server which handles the application protocol and maintains incoming connections from clients. The latter is the adaptation layer which exchanges request/response data and communicates with WSN at the back-end of the web server.

Wisegate-P/S is a monitoring model for publish and subscribe scheme for periodical sensor data acquisition from WSN. The Wisegate web server is mutated to subscribe clients that are introduced to the server to sensor services. The server gathers periodically generated sensor data from WSN to its database at the back-end and publishes these data to clients according to subscribed topics (i.e., sensor data request contracts of clients). A topic is recorded in the server per client subscription.

4.1 Arhitectural Design

Wisegate-P/S has multiple components as illustrated in Fig. 1. A user agent interacting with the client application, subscribes his/her to the server and publishes the content according to his/her topic with a response message. A sensor agent operating at the back-end, connects to a sensor node via the websocket interface and gathers data from sensor node connected to itself. This agent saves sensor data in a database on the server. The topics and search engine records the topics of the subscribed clients in this database. When a sensor agent saves data, the topics and search engine looks up client topics in the database, finds user agents which request this sensor data and sends it to those user agents.

The user agent module has several sub modules as shown in Fig. 2a. The subscribe topic parser divides the subscribe message to topic tokens and sends them to its topic recorder sub-module. Topic recorder saves the tokens to the database. These sub modules perform their operation once when a client connects to the server for subscription. Message publisher gets sensor data coming from the search engine, generates the client message and sends it to the client.

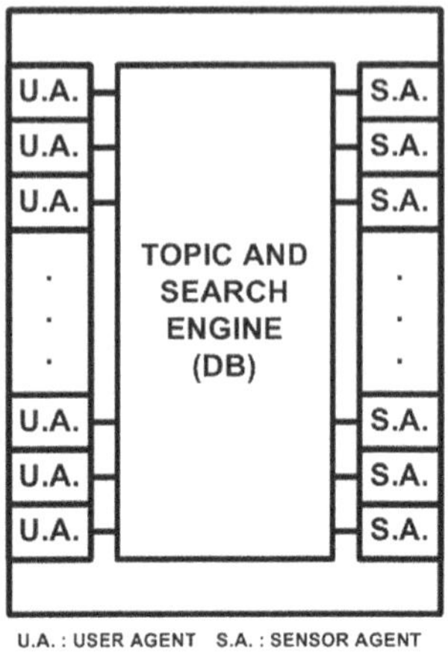

Fig. 1 Operational parts in WiSEGATE P/S

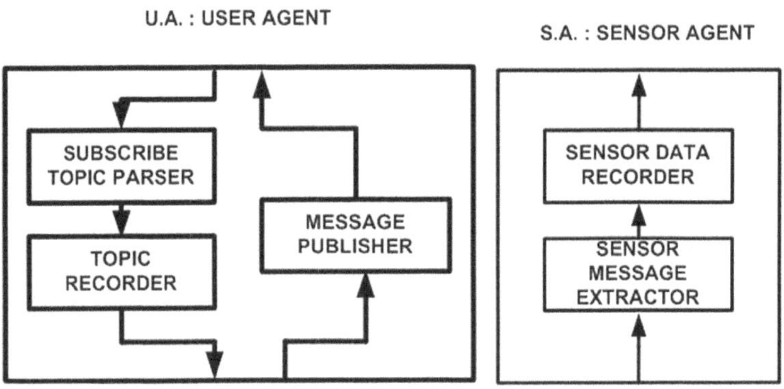

Fig. 2 Modules in user and sensor agent

The sensor agent module has two sub modules as illustrated in Fig. 2b. The sensor message extractor gets incoming sensor data from the websocket connected to the sensor node and sends it to the sensor data recorder. Sensor data recorder updates sensor data of the corresponding sensor node and sensor type.

4.2 *Investigation, Subscription and Data Model*

Wisegate-P/S server should listen all sensor nodes and investigate the services provided by those nodes before a publish/subscribe operation can succeed. So, after a connection of a sensor node is done as illustrated in the sequence diagram depicted in Fig. 3, server sends a service investigation message to it. The sensor node replies

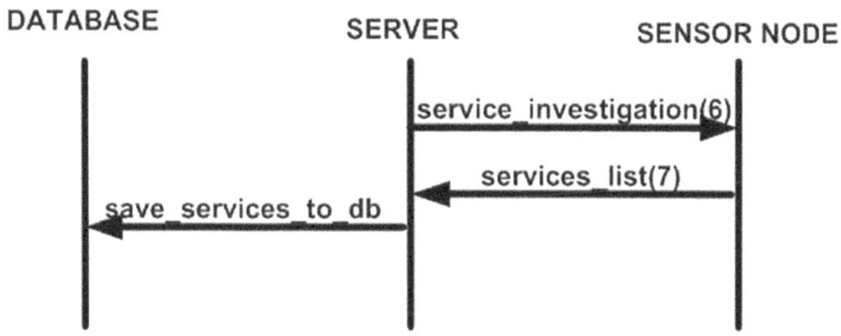

Fig. 3 Service investigation operation

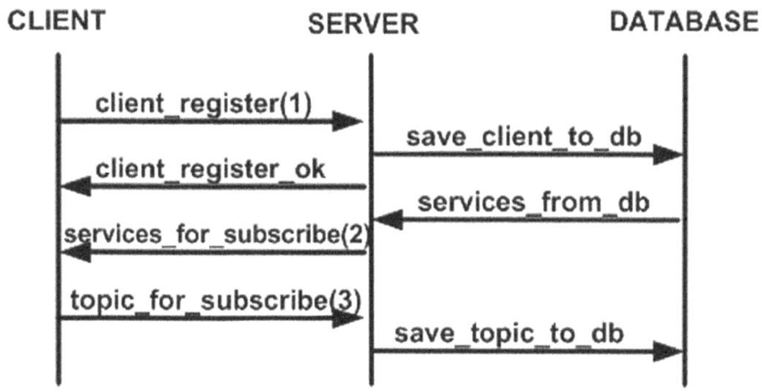

Fig. 4 Register operation of a client

with the services provided by itself to the server. The services list message structure is given in Fig. 7. When server obtains the services list, it saves the services and the sensor node id to the database and creates a sensor agent for this sensor node over the established connection.

A client should register to the server following the steps as illustrated in a sequence diagram given in Fig. 4 before subscription. At the beginning, client sends a username and a password to the server to register. After registration, server sends a set of possible sensor services introduced to itself by sensor nodes within a response. The structure of this message is given in Fig. 7. After that, client can subscribe to the server. In subscription process, client creates his/her topic with services obtained in the message explained above. The message structure of a topic is given in Fig. 7. Client sends the topic message to the server including the userid obtained after registration operation. After that, server reads the topic and save the topic of a client in the database. After subscription, client does not need to send his/her topic in the following login operations.

Client login process is done with several steps as illustrated in the sequence diagram in Fig. 5 Whenever a client wants to login to the server, s/he should provide a user name and password which is recorded in a register operation to use the services determined in his/her topics. For login, server subscribes the client and creates an user agent for publish operation.

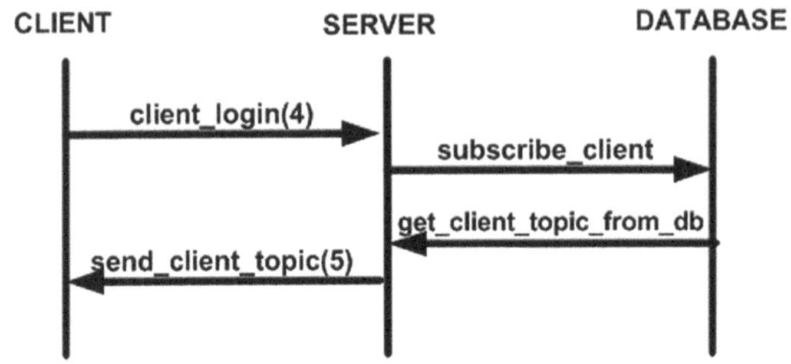

Fig. 5 Login operation of a client

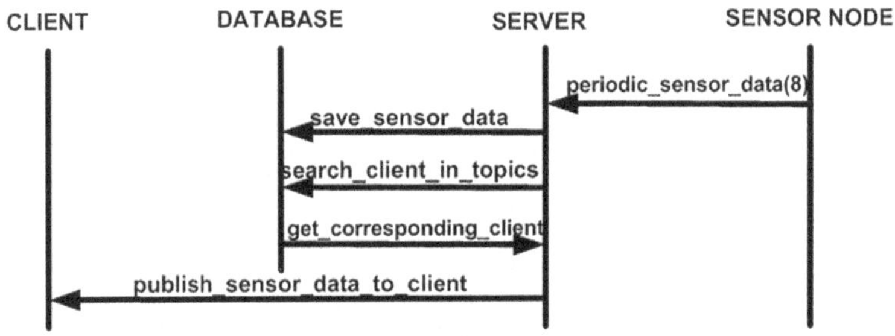

Fig. 6 Publish data

(1) client_register_message={client_name,client_password}
(2) services_for_subscribe_message={services:[service]}
service={node_id,service_id,service_description,possible_publish_periods:[int]}
(3) topic_for_subscribe_message={client_id,topic_parts:[topic_part]}
topic_part={node_id,service_id,publish_period}
(4)client_login_message={client_name,client_password}
(5)send_client_topic_message={client_id, topic_parts:[topic_part]}
topic_part={node_id,service_id,service_period}
(6)service_investigation_message={service_start}
(7)services_list={services:[service]}
service={service_type,service_description}
(8)periodic_sensor_data_message={sensor_id,service_type,value}
(9)publish_message={sensor_id,service_type,value}

Fig. 7 Message and data types

Publish operation is depicted in Fig. 6. Whenever a sensor node prepares its data, it sends it to the corresponding sensor agent of server in a periodic sensor data message with a structure depicted in Fig. 7. Server reads the sensor value, node id and service type in the message and stores it in database. After that, server searches topics and finds the corresponding clients and send the sensor data using the corresponding user agent. Then, user agent publishes the data to the client.

All data and message formats are depicted in Fig. 7. Server interacts with client and sensor nodes with messages in JSON-like data structures. For topics and search engine, server provides a NoSQL database and saves sensor data and topics of clients in it. Database saves a collection of sensor data with tuples including node ID, sensor type, sensor data and time stamp values. It saves collection of client topics with tuples including unique generated client ID, sensor node ID and sensor type.

5 Experimental Setup

We have implemented the Wisegate architecture and its P/S services to get more insights into its capacity of establishing and maintaining connections. We have analyzed the performance of the system in terms of end-to-end latency and response time characteristics for varying number of connections. We developed our software on a custom sensor node (6LN) and border router (6LPR) as explained below. Since it is aimed to reduce the power consumption of the sensor network, 6LoWPAN technology is preferred as the access layer protocol. 6LoWPAN adaptation layer allows IPv6 packets to be carried efficiently within small link layer frames, such as those defined by IEEE 802.15.4 [13]. 6LoWPAN is an open standard defined in RFC 6282 by the Internet Engineering Task Force (IETF), as the full name implies IPv6 over Low-Power Wireless Personal Area Networks. Today, many sensor operating systems such as Contiki, TinyOS, RiOT, OpenWSN, Nano-RK, and so on, support the 6LoWPAN protocol. Our software stack is built on Contiki Operating System [14] which is a standard platform for many IoT applications. The Contiki software has a strong protocol stack and it is regularly updated to support common IoT standards. It is also compatible with TinyOS, and has the advantages of the operating system's application software interface. In addition, Contiki OS has been developed for microcontrollers with small memory; i.e. basic Contiki configuration requires only 2 KB of RAM and 40 KB of ROM. However, depending on the software libraries and applications used, RAM requirement may be in the order of 10 s of KByte.

5.1 IoT Node—6LN

Sensortag (CC2650STK) hardware developed by Texas Instruments is used as the sensor node in our experimental setup. The sensor node (6LN) which is illustrated in Fig. 8 has a coin cell battery holder, various sensors on board for different applications like light, temperature, pressure, sound, vibration, magnetic field measurements. We built our WSN on contiki [15] operating system and its own SICSlowpan (6LoWPAN) stack implementation.

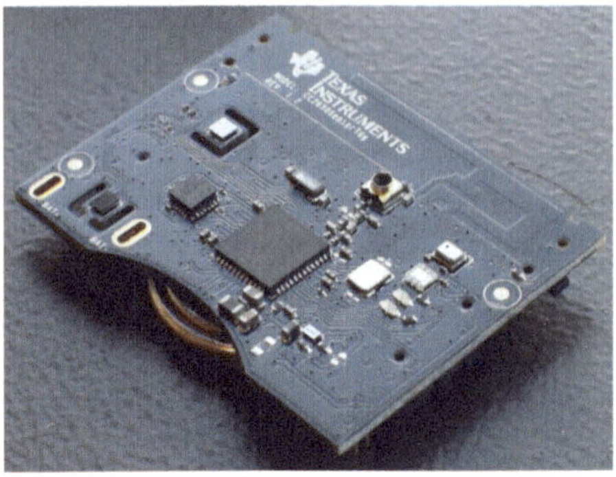

Fig. 8 CC2650STK-sensortag

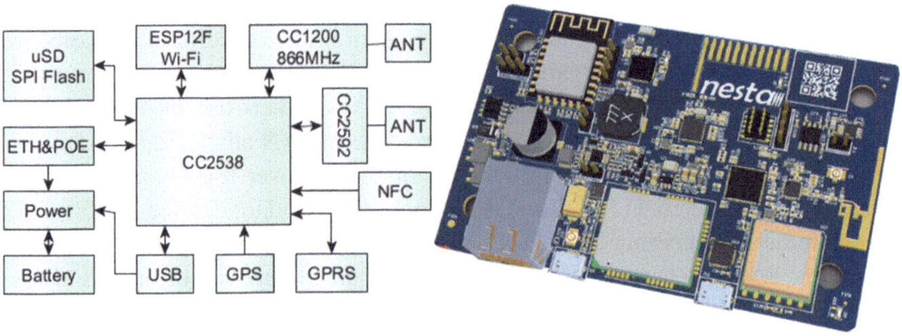

Fig. 9 Border router-6LBR

5.2 6LoWPAN Border Router—6LBR

In our setup, 6LoWPAN network, LPGv1.1 developed by Nesta Electronics is used as the border router. The hardware components of the Border Router and the corresponding symbolic model are shown in Fig. 9. The hardware has 2.4 GHz CC2538 SoC and 866 MHz CC1200 transceiver to create and manage a 6LoWPAN network. On the IPv4 side with SPI-based ethernet PHY and enc28j60 with MAC. There is a power circuit that can be powered from battery, via Ethernet (Power over Ethernet—PoE) or USB, and can perform the necessary switching operations. There are also optional Wi-Fi, GPRS and NFC interfaces that are still under development.

As the Border router software, 6LBR [17] project was used and some changes were made for hardware differentiation. NAT64, DN64 running on the border router is used for the nodes to connect Internet or to transmit data on the local network. The protocol stack configurations running on 6LBR and 6LN are shown in Fig. 10.

The block diagram showing the connections for the entire system is given in Fig. 11. 6LNs connect to the server running Wisegate via 6LBR with IPv4. Since

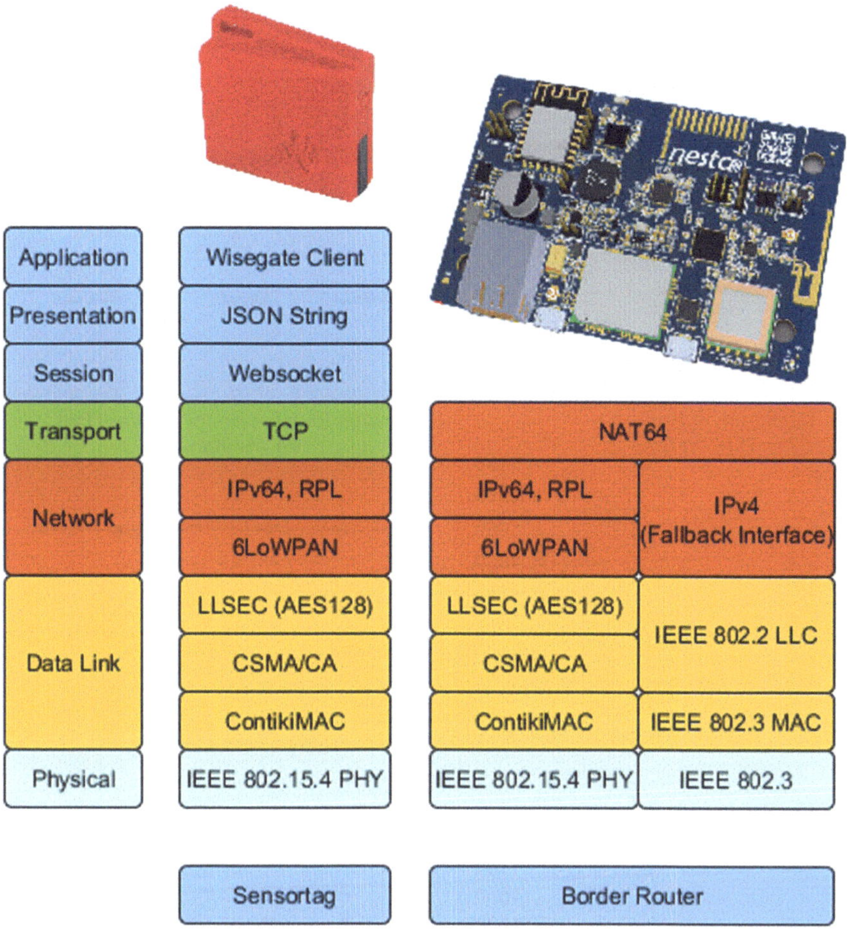

Fig. 10 6LN and 6LBR protocol stack

the Wisegate application is on the same local network as the 6LoWPAN network, it is not necessary to support IP64 devices, however, using IP64 allows Wisegate to run from a server on the Internet.

6 Results and Conclusions

We devised experiments to connect N web clients to 1 Sensor Node (N-1) over the Internet using the setup defined in Sect. 5. In the experiments, number of clients (N) is increased in logarithmic scale from 1 up to 32. In each experiment, clients are subscribed to simultaneously monitor a sensor node managed by WiSEGATE P/S. The sensor node senses the the environment periodically with a sensing period (SP)

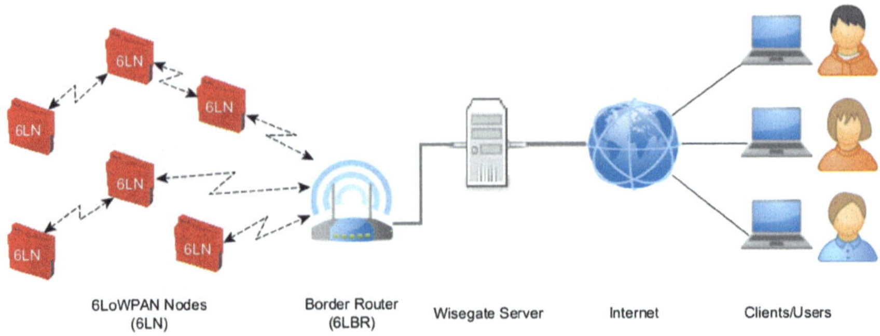

Fig. 11 6LN, 6LBR, Wisegate connections

Fig. 12 Results of N is up to 32 and SP 5000 ms

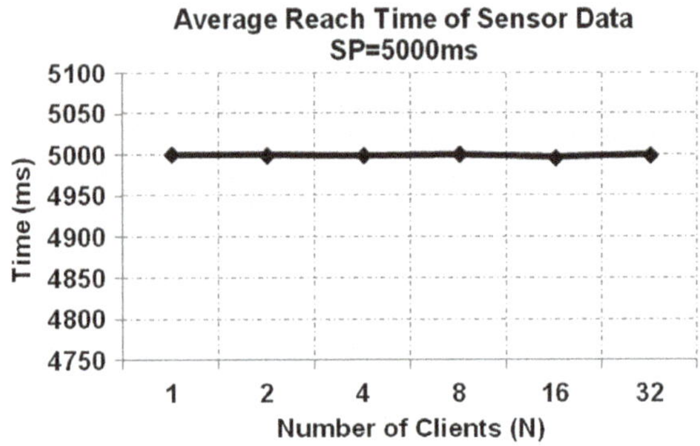

Fig. 13 Results of N is up to 32 and SP 1000 ms

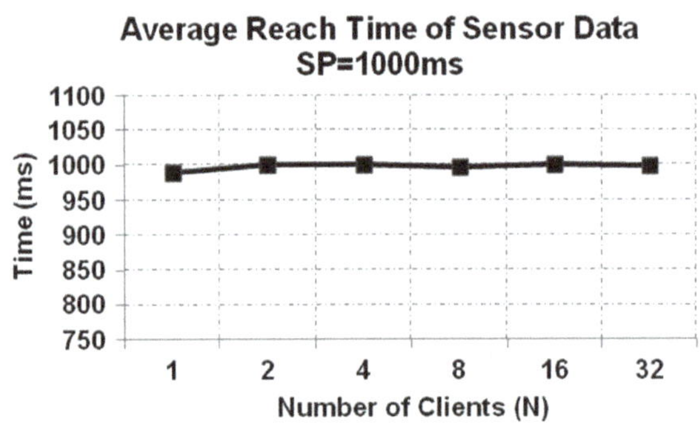

and publishes sensor data to clients. In each client, a program is used to obtain sensor data and get differences between arrival time of two consecutive sensor data. We call this value as D-value throughout the evaluation of results. In each experiments, we get 50 consecutive D-values from each client. We repeat each experiment 10 times and calculate the averages of D-values. Figures 12, 13, 14 and 15 illustrate the average D-values with different SPs of sensor node. Because of jitter in underlying MAC in both client-server and server-sensor node side, D-values can vary above and

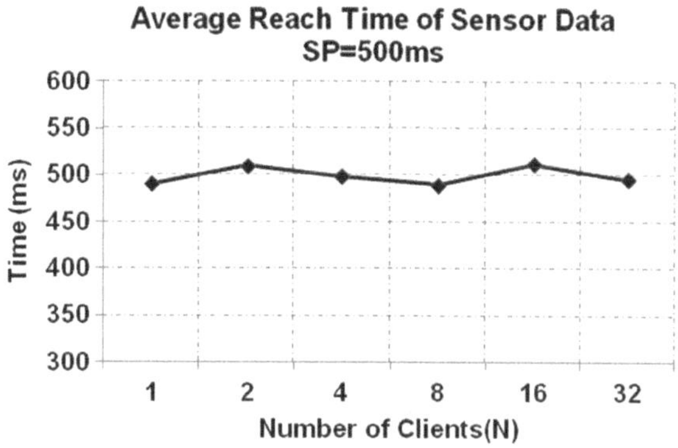

Fig. 14 Results of N is up to 32 and SP 500 ms

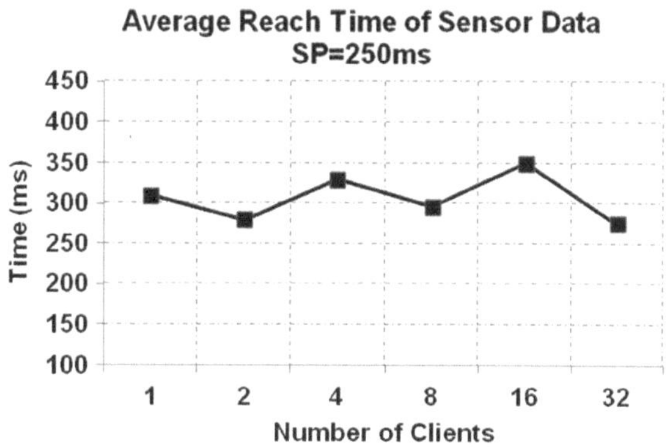

Fig. 15 Results of N is up to 32 and SP 250 ms

under of SP. Even the existence of variation in D-values, we observe that when SPs are 5000, 1000 and 500 ms, the average of D-values are close to SP. In addition, we observe that increasing number of client does not affect the change in average of D-values. However, when we set SP to 250 ms, the averages become larger than SP; but still with considerably low jitter. This results indicate a good limit for up to 32 clients reaching the same sensor simultaneously over the Internet without any real-time delay. As a summary, (N-1) test results are very promising for different number of clients and sampling rates. We plan to conduct more experiments to analyze the limits of our model for (1-N) and (N-N) connections as well as more complex WSN structures as a future work. We also intend to design additional services over Wisegate P/S and offer it as an open service platform for IoT applications.

References

1. K.C. Serdaroglu, S. Baydere, WiSEGATE: wireless sensor network gateway framework for internet of things. Wirel. Netw. **22**, 1475–1491 (2016)
2. J.J.P.C. Rodrigues, P.A.C.S. Neves, A survey on IP-based wireless sensor network solutions. Int. J. Commun. Syst. **23**(8) (2010)

3. H. Ting, C. Xiaoyan, Y. Yan, A new interconnection scheme for WSN and IPv6-based, in *Proceedings of Information, Computing and Telecommunication 2009 Conference, YC-ICT'09*, 2009
4. Y. Chen, A smart gateway design for WSN health care system, Master thesis report, 2009
5. A. Santos, J. Macedo, A. Costa, M. Joäo Nicolau, Internet of things and smart objects for M-health monitoring and control. Procedia Technol. J. Elsevier, **16**, 1351–1360 (2014)
6. M. Harvan, Connecting wireless sensor networks to the internet: a 6lowpan implementation for TinyOS 2.0, Master thesis report, School of Enginnering and Science, Jacobs University Bremen, 2007
7. TinyOS overview (2017), http://tinyos.stanford.edu/tinyos-wiki/index.php/TinyOS_Overview
8. G. Montenegro, N. Kushalnagar, J. Hui, D. Culler, Transmission of IPv6 packets over IEEE 802.15.4 networks. RFC 4944, IETF, 2007
9. A. Dunkels, Mmspeed: full TCP/IP for 8-bit architectures, in *Proceedings of the 1st International Conference on Mobile Systems, Applications and Services, MobiSys 03* (2003)
10. C. Bormann, A.P. Castellani, Z. Shelby, CoAP: an application protocol for billions of tiny internet Nodes. IEEE Int. Comput. **16**(2) 2012
11. MQTT (2017), http://mqtt.org
12. I. Fette, A. Melnikov, The web socket protocol. RFC 6455, IETF (2011)
13. Z. Shelby, C. Bormann, *6LoWPAN: The Wireless Embedded Internet*, vol. 43 (Wiley, 2011)
14. T. Borgohain, U. Kumar, S. Sanyal, Survey of operating systems for the IoT environment (2015), arXiv:1504.02517
15. A. Dunkels, B. Gronvall, T. Voigt, Contiki-a lightweight and flexible operating system for tiny networked sensors, in *Proceedings of the 29th Annual IEEE International Conference on Local Computer Networks 2004*, pp. 455–462, Nov 2004
16. B. Tudosoiu, Feasibility Study: Minimum Viable Device to Support Internet of Things Realization, Mobile Heights, (Lund, Jan 2014)
17. L. Deru, S. Dawans, M. Ocaña, B. Quoitin, O. Bonaventure, redundant border routers for mission-critical 6lowpan networks, in *Real-World Wireless Sensor Networks* (Springer International Publishing, 2014) pp. 195–203

Toward Data Governance in the Internet of Things

Sabrina Sicari, Alessandra Rizzardi, Cinzia Cappiello, Daniele Miorandi and Alberto Coen-Porisini

Abstract The diffusion of Internet of Things (IoT) technologies not only enables the provision of advanced and valuable services, but also raises several challenges. First of all, the increasing number of heterogeneous interconnected devices creates scalability and interoperability issues, and thus, a flexible middleware platform is needed to manage all the sources together with all the tasks related to data collection and integration. In fact, the large amount of data has to be properly managed. In particular, on the one hand, data have to be protected from security threats; on the other hand, it is necessary to consider that data are useful only if their quality is suitable for the processes in which they have to be used. For these reasons, it is important that applications/users that aim to exploit the collected data are aware of data quality and security levels in order to understand if data can be trusted and thus used. In this chapter, we present a distributed architecture for managing IoT data extraction and processing that also includes algorithms for the assessment of data quality and security levels of considered sources. A prototype of such an architecture has been realized; through a user interface, it is possible to access data services able to filter data from IoT devices on the basis of security and data quality requirements. The chapter describes the prototype and shows some experiments performed by using several real-time open data feeds characterized by different levels of reliability, quality and security.

S. Sicari · A. Rizzardi · A. Coen-Porisini
Dipartimento di Scienze Teoriche e Applicate, Università degli Studi dell'Insubria, via Mazzini 5, 21100 Varese, Italy
e-mail: sabrina.sicari@uninsubria.it

A. Rizzardi
e-mail: alessandra.rizzardi@uninsubria.it

A. Coen-Porisini
e-mail: alberto.coenporisini@uninsubria.it

C. Cappiello (✉)
Politecnico di Milano, Piazza Leonardo da Vinci 32, 20133 Milano, Italy
e-mail: cinzia.cappiello@polimi.it

D. Miorandi
U-Hopper, via A. da Trento 8/2, 38122 Trento, Italy
e-mail: daniele.miorandi@u-hopper.com

R.R. Yager and J. Pascual Espada (eds.), *New Advances in the Internet of Things*, Studies in Computational Intelligence 715, DOI 10.1007/978-3-319-58190-3_4

Keywords Internet of Things · Security · Data quality · Middleware

1 Introduction

The diffusion of Internet of Things (IoT) technologies turns everyday objects into *Smart Objects*. A global network infrastructure allows such physical objects or *things* to interact among themselves and with the environment where they are placed, in order to fulfill a given goal [1]. Such an interaction, together with the collection and integration of sensed data, not only enables the provision of innovative and customized services to individuals and businesses in different application domains, but also raises many challenges. In fact, the resulting system may include an extremely large number of heterogeneous devices and thus creates scalability and interoperability issues. Hence, it is necessary to deal with the variety of protocols, domains and applications, and with the fact that estimations state in 2020 the number of Internet-connected things will reach 20 billions. Moreover, connecting physical objects to the Internet implies the transfer and management of a high amount of data that need to be properly governed. In particular, in this chapter, we aim to highlight that data governance concerns the analysis of data processes and risk management and, in particular, it includes the assurance of regulatory compliance, security, privacy and quality.

Security and privacy are widely acknowledged to represent critical issues in such a context [2]. Besides security issues related to the communications over the Internet, it is also necessary to consider that devices have limited interfaces to monitor network intrusion, and they have data vulnerabilities that can often be exploited to attack the system. In this scenario, the confidentiality and the integrity of the transmitted and stored information has to be guaranteed, and authentication and authorization mechanisms have to be provided to prevent unauthorized users or devices to improperly access the system. Furthermore, taking into account that data are often related to personal and/or sensitive information, privacy of users, intended as the ability to support data protection and users anonymity, has to be ensured [3].

As stated above, the large amount of available data enables the design of smart services. However, in order to obtain valuable results, it is necessary to consider that not all the values might be relevant: errors, missing or outdated values can negatively affect decisions [4, 5]. Data quality represents another essential requirement for the adoption at scale of IoT services; provided results should be correct and reliable or at least the users should be aware of both security and quality level of the data being accessed, in order to take informed decisions about their usage.

For these reasons, we claim that an effective data governance should be supported by a system able to manage heterogeneous data sources and to evaluate the security and the quality of the information being collected, processed and transmitted, possibly in real-time and in an automatic manner. The design of such a system has also to deal with the dynamism of the IoT environments and with the fact that the composition of input sources can evolve over time adding new sources or eliminating old ones. In this chapter, we present novel algorithms for evaluating the data security and quality levels of the data provided by different sources over time.

Such algorithms are integrated in an existing IoT middleware, named *NetwOrked Smart objects* (*NOS*) presented in our previous work [6–8]. NOSs are computationally powerful devices connected to create a distributed processing and storage layer, able to process the data gathered from IoT data sources. NOSs collect the data generated by nearby IoT devices, process them and finally transmit the processed data on a publish/subscribe broker. Such a middleware includes functionalities for users and applications to dynamically specify their requirements in terms of data security and quality levels. In this way, the architecture is able to assess data security and quality metadata and filter out only the data that satisfy users/applications needs. This adaptive behavior represents a clear innovation over conventional one-size-fits-all approaches, which provide the same information to all consumers without considering their specific preferences. We also provide a prototype of the illustrated architecture in which real-time open data feeds are used.

The chapter is organized as follows: Sect. 2 describes literature contributions that consider security and data quality issues in IoT scenarios. Sections 3–5 present, respectively, the architecture and the algorithms for evaluating security and data quality levels. The implemented prototype is instead described in Sect. 6, along with the results obtained by the validation phase. Section 7 ends the paper.

2 Related Work

One of the main factors limiting the growth and take-up of IoT is the lack of a reference model [9]. There have been many projects that tried to design a common architecture based on the analysis of the needs in this scenario. For example, an architectural reference model for the interoperability of IoT systems has been the main goal of the Internet of Things Architecture (IoT-A) project [10]. A dynamic architecture for services orchestration and self-adaptation has been proposed in Internet of Things Environment for Service Creation and Testing (IoT.EST) [11]. The project defines a dynamic service creation environment that gathers and exploits data from sensors and actuators making use of different communication technologies and formats. Such an architecture deals with issues such as the composition of business services based on reusable IoT service components, the automated configuration and testing of services for "things" and the abstraction of the heterogeneity of underlying technologies to ensure interoperability. Also the FP7 COMPOSE (Collaborative Open Market to Place Objects at your Service) project [12] focuses on composition. It aims to design and develop an open marketplace for IoT data and services. The basic concept underpinning such an approach is to treat smart objects as services, which can be managed using standard service-oriented computing approaches and can be dynamically composed to provide value-added applications to end users. Another architecture has been proposed in the Ebbits project [13], that designed a SOA platform based on open protocols and middleware, effectively transforming IoT subsystems or devices into web services with semantic resolution. The goal was to allow businesses to integrate IoT into mainstream enterprise systems and to support interoperable end-to-end business applications.

Considering all these contributions, it is possible to state that researchers only agree that the basic model is a 3-layer architecture composed of application, network and perception layers [9, 14]. However, in the literature, several contributions propose other models that add more abstraction to the IoT architecture [14, 15].

In our work, we also tried to add different abstraction layers for designing a lightweight and flexible middleware for IoT applications [8]. In this chapter, we aim to highlight the novelty and relevance of this solution, since it is the unique architecture that provides a comprehensive approach for managing data gathered from heterogeneous sources both addressing security and data quality issues. In fact, in the literature, security and data quality issues have been addressed, but separately, as discussed in the next sections.

2.1 Security Issues in IoT

In some proposal, security issues have been addressed. Typically, they focus on data communications; they enforce data to be exchanged according to strict protection constraints considering the heterogeneity of devices and communication technologies. Indeed, devices can be characterized by different technologies; for example, many smart devices can natively support IPv6 communications [16, 17], while other existing deployments might not support the IP protocol within the local area scope and this requires the design of ad hoc gateways and middlewares [18]. Relevant contributions on security oriented IoT middlewares include as follows: VIRTUS [19], which relies on the open extensible Messaging and Presence Protocol (XMPP) to provide secure event-driven communications; Otsopack [20] and naming, addressing and profile server (NAPS) [21], which are data-centric frameworks based on the usage of HTTP and representational state transfer (REST) interfaces.

Security aspects were also the central points of projects such as uTRUSTit [22] and Butler [23]. The approach pursued in the former one is to directly integrate the user in the trust chain, guaranteeing transparency in the underlying IoT security and reliability properties. If successful, the uTRUSTit approach shall enable system manufacturers and system integrators to express the underlying security concepts to users in a comprehensible way, allowing them to reason on the trustworthiness of such systems. Butler aims to allow users to manage their distributed profile by allowing data duplication and identity control over different applications. The final purpose is to deliver a framework able to dynamically integrate user data (e.g., location, behavior) in privacy and security protocols.

2.2 *Data Quality Issues in IoT*

The huge number of data sources in IoT is considered a positive aspect for data fusion and for the extraction and provisioning of advanced services. However, it is important to exploit the benefits provided by the quantity of data only if a certain quality is guaranteed. Several literature contributions recognize data quality as one instrumental issue in IoT research. A survey was recently dedicated to this topic, stating that data quality is the basis for sound decisions [24]. In the survey, authors consider work that addressed quality issues in data streams and RFID data. They show that the main data quality dimensions are accuracy, confidence completeness, data volume and timeliness.

Other additional data quality dimensions to consider are mostly related to the infrastructure through which data are offered; in fact, they include the ease of access, the access security and the availability. Also in [25], quality is cited as a key parameter for an efficient access, and use of IoT data and services, and [26] claims the need of control over data sources to ensure their validity, information accuracy and credibility. Other literature contributions are focused on specific data quality aspects and issues. Metzger et al. [27] addresses only data accuracy timeliness and the trustworthiness of the data provider. In particular, authors propose anomaly detection techniques to remove noise and inaccurate data in order to improve data quality.

New data quality dimensions have been introduced in [28]. In fact, authors focus on uncertainty, redundancy, ambiguity and inconsistency. Such dimensions are mainly related to the fact that, in an IoT environment, data may be gathered from different sources, and they can be characterized by different precision or accuracy or they can monitor the same phenomenon reporting duplicates or inconsistent values. All these issues have a negative effect on the source reliability.

3 Networked Smart Objects Architecture

The architecture called NOS, proposed in [8], is illustrated in Fig. 1. It provides interfaces for the interaction with the sources and with the users. In fact, on the one hand, the architecture gathers input data from different kinds of sources (the so-called *E-Nodes*) that are represented by heterogeneous devices (e.g., wireless sensor networks, RFID, NFC, actuators, social networks); on the other hand, it lets the users access the offered IoT-based services through Internet-connected mobile devices (e.g., smartphone and tablet). In the following, we briefly describe how the architecture works.

Starting from the sources, the architecture provides a service for the source registration by means of HTTP protocol. Registered sources are associated with an identifier, and, optionally, with a geographical position and/or an encryption scheme, including the proper keys for interactions with NOSs. For each incoming data, NOSs extract the following information: (i) the data source, which describes the type of node (i.e., the identifier in case of a registered source); (ii) the communication mode,

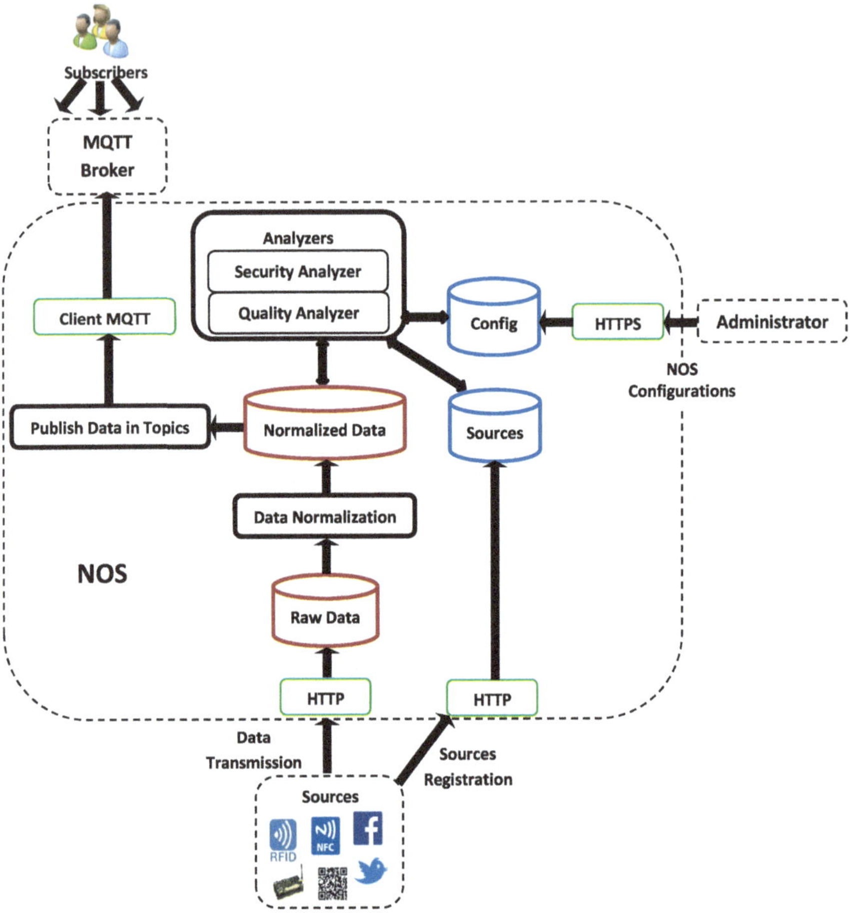

Fig. 1 NOS architecture

that is the way in which the data are transmitted (e.g., discrete or streaming communication); (iii) the data schema, which represents the type (e.g., number and text) and the format of the transmitted data; (iv) the metadata describing the data content; (v) a timestamp describing when the data were received by NOS. The HTTP communication protocol is also used among NOSs and the data sources for the data transmission. Since received data are highly heterogeneous, they are stored in the *Raw Data* repository, and, periodically, they are processed by the *Data Normalization* and *Analyzers* modules. The former puts the data in the format specified in Fig. 2. The latter periodically extracts the normalized data and computes the relevant security and data quality indicators (the related algorithms are described in Sects. 4 and 5).

The processed data are used for providing services to the target users. The user interface is based on the Message Queue Telemetry Transport (MQTT) protocol, that

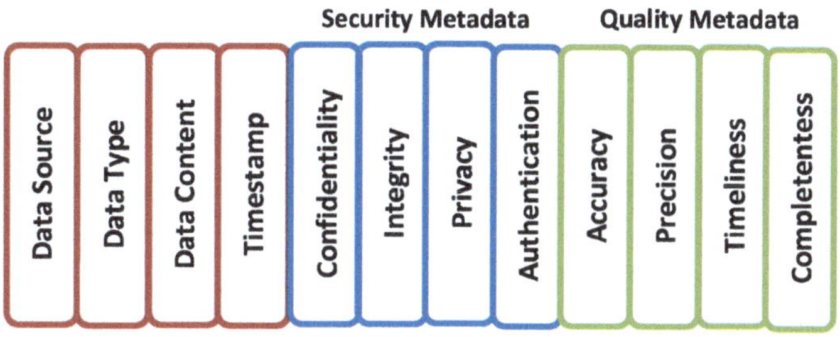

Fig. 2 NOS data format

is a lightweight publish/subscribe protocol [29] specifically designed for resource-constrained devices. The MQTT client exchanges messages with the MQTT broker by means of publications and subscriptions to topics. Such a mechanism is adopted to support interactions between services and IoT devices. The architecture includes a module in charge of assigning data items to the corresponding topic and to publish them on a MQTT broker, as depicted in Fig. 1. For this task, a taxonomy is needed: for example, for publishing temperature information of a sensor with identifier *sensorId* could be *sensor/temperature/sensorId*. Note that subscribers may register for specific topics at runtime, and our architecture provides a mechanism for dynamic subscription and unsubscription to topics. According to the MQTT protocol, messages can be published with a quality of service (QoS) parameter indicating that a message should be delivered "at most once," "at least once" and "exactly once." MQTT also supports persistence of messages to be delivered to future clients that subscribe to a topic, and may be configured to send messages of specific topics when the subscriber connection is abruptly closed. These parameters are specified in the *Config* storage unit. Summarizing, a typical MQTT message includes the following parameters: (i) the topic; (ii) the data value; (iii) the QoS level; and (iv) the retain value.

4 Data Quality Evaluation

IoT architectures are designed to handle streams of data gathered from millions of intelligent devices; therefore, new challenges in data quality assessment raise. In fact, data volume, velocity and variety need to be properly addressed. The first two issues can be handled by considering a window-based approach for which the data quality of the data streams results from the periodic assessment of the set of values included in the different time windows. Different windows can be rapidly analyzed by using parallel distributed processing (e.g., map-reduce approaches). Data variety requires adaptive mechanisms, able to activate the appropriate data quality dimension and the related assessment metric. For example, in case of numbers (e.g., temperature values gathered from a specific sensor or provided by a sensor), accuracy and precision have to be calculated, but in case of text (e.g., tweets), the source reputation could be more relevant than the evaluation of a syntactic accuracy. However, all the data quality metadata assessed for the different windows are aggregated to define the source data quality level. In particular, in our approach, the data quality metadata have been defined as scores in the range [0, 1]. For the case study defined in Sect. 6, timeliness, completeness, accuracy and precision levels have been considered [30, 31] and evaluated by the *Quality Analyzer* (see Fig. 1).

Timeliness is defined as the temporal validity of data and is calculated on the basis of the freshness of data and on the frequency of data updates. The former is called *Currency* and defines the interval from the time when the value was sampled to the time instant at which data are received by NOSs. The latter is called *Volatility* and indicates the amount of time units (e.g., seconds) during which a data remains valid; it is usually associated with the type of phenomenon that the system has to monitor and depends on the timescale of its dynamics.

Completeness provides information about the quantity of values received by the source. In particular, it analyzes if the data set includes all the values that a sensor or device was supposed to gather. In fact, it is calculated as the amount of collected values in a given time interval with respect to the amount of expected values. Completeness is important since the percentage of missing values can be a good indicator for revealing sensors inefficiencies or communication issues.

Accuracy is usually defined as correctness, and it is measured as the degree of similarity between the value stored in the system and the right value. More formally, the accuracy is based on the evaluation of the error ε_{acc} resulting from the difference between the sensed value v_n and a reference value v_{ref}. Accuracy is usually related to *Precision*, conceived as the degree to which further measurements of the same phenomenon in a close time instant provides the same or similar results. Precision is often specified in terms of the standard deviation of the measured values: the smaller the standard deviation, the higher the precision. A correct representation is characterized by accurate and precise values. However, in continuous quality monitoring, changes in accuracy and precision can reveal errors or changes in the monitored process. In particular, precise but inaccurate values can be caused by changes in the monitored phenomenon or by faulty sensors [30].

5 Security Evaluation

Together with data quality metadata, security metadata are exploited to understand the nature and reliability of the sources managed by the IoT platform. The *Security Analyzer* is responsible for the security assessment and must be able to access the *Sources* storage unit in order to analyze the received data in relation to the source that sent them to NOSs.

In more detail, NOSs associate a score in the range [0, 1] to each security metric. As in the IoT context sensitive data are often managed, the security scores are intended as levels of *confidentiality* and *integrity* of the received information, *pri-*

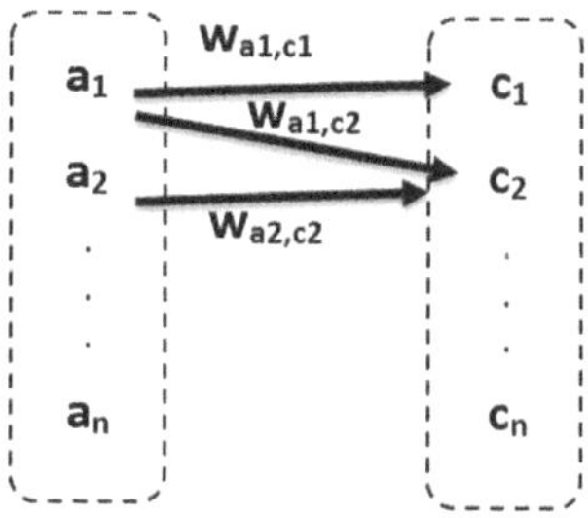

Fig. 3 Weighed relationships among attacks and countermeasures

vacy of the transmitting source and *authentication* (i.e., the robustness of the source authentication). Note that malicious devices may be represented by non-registered sources, which send violated data to the IoT platform or execute malicious actions toward those transmitted by non-malicious ones (e.g., spoofing and sniffing).

In order to assess security, it is necessary to consider two groups of elements: a set A of threats/attacks a_n and a set C of security countermeasures c_m. The former includes the attacks that may impact on the data managed by the platform (e.g., data violation, unauthorized access, masking and impersonation). The latter regards the countermeasures available in the platform to face the attacks (i.e., encryption, authentication and key pre-distribution). The security model considered by this algorithm links the attacks of a_n with the corresponding countermeasures in c_m. The taxonomy of the security attacks and the related countermeasures is retrieved from [32], which also considers that each countermeasure is characterized by a degree of resistance to a violation or to an attack attempt. On such a basis, we are able to define the relationships among attacks and countermeasures, considering that an attack can be tackled through a plurality of countermeasures and a countermeasure can face more than one attack. Each relationship is associated with a weight w_{a_n,c_m} in the range $[0 : 1]$, which represents the level of robustness of the countermeasure c_m with respect to the attack a_n (see Fig. 3).

The identified relationships can be classified on the basis of the considered security metrics in a way to obtain four groups that might also overlap: (i) g_{conf} for attacks-countermeasures related to the data confidentiality; (ii) g_{int} for the pairs related to data integrity; (iii) g_{pri} for privacy issues; (iv) g_{auth} for the pairs concerning sources' authentication. Note that such a model has to be defined at design time, stored in the collection named *Config* and can be updated at runtime when new attacks and/or countermeasures have to be considered.

Table 1 shows some examples of attack-countermeasure pairs derived from the used taxonomy and classified in the groups described above.

The weights associated with the relationships are at the beginning set to 1. They might be updated at runtime on the basis of the malicious events that occur in the IoT system (detected by a monitoring system installed on NOSs). Weights can thus vary over time in a dynamic way; such a process of automatic adjustment is performed by means of a well-known learning approach, namely difference temporal learning [33]. Please refer to [8] for further details on this method.

Table 1 pairs attack-countermeasure

Attack	Countermeasure	Group
(1) Packet sniffing	Data content encryption	g_{conf}
(2) Password attack	Complex password generation	g_{conf}, g_{auth}
(3) Man-in-the-middle attack	Data content encryption	g_{int}
(4) Session hijacking attack	Secure session establishment	g_{int}, g_{auth}
(5) Identity spoofing	Identity encryption	g_{pri}
(6) Key impairment	Secure key distribution scheme	g_{conf}, g_{auth}

Once the attacks/countermeasures model is defined, the algorithm computes for each incoming data the related security scores on the basis of the actual weights and of the data source s_k. A general equation for the assessment of the different security dimensions sec_{dim} (i.e., confidentiality (sec_{conf}), integrity (sec_{int}), privacy (sec_{pri}) and authentication (sec_{auth})) can be defined as:

$$sec_{dim} = \frac{|A_{dim,s_k}|}{|A_{g_{dim}}|} \cdot \frac{\sum_{i \in A_{dim,s_k} j \in C_{dim,s_k}} w_{i,j}}{|C_{dim,s_k}|} \quad (1)$$

where: A_{dim,s_k} is the number of attacks related to a specific security dimension that the source s_k could suffer; $A_{g_{dim}}$ is the total number of attacks included in the group g_{dim} (valid for any type of sources); C_{dim,s_k} is the number of countermeasures adopted by s_k related to the attacks to the specific security dimension included in A_{dim,s_k}. The sum of the weights considers only the weights between the attacks in A_{dim,s_k} and the countermeasures in C_{dim,s_k}.

For example, let us consider confidentiality, and let us suppose that the source s_k adopts AES for encrypting its data; moreover, it also adopts a 8-bit length password as credential for ensuring both confidentiality and authentication. As shown in Table 1 (points 1 and 2), AES is a countermeasure associated to the g_{conf} group, while the password is associated to both g_{conf} and g_{auth} groups. The steps performed by NOS to assess over the time the confidentiality score sec_{conf} are the following:

- The initial weights corresponding to the two pairs attack-countermeasure (i.e., AES-packet sniffing, 8-bit password-credential violation) are set to 1, and the first confidentiality score sec_{conf} is evaluated;
- During the system operations, the platform recognizes no violated packets from the source s_k, but several times its password has been intercepted (e.g., through brute-force attack). As a consequence, the weight related to the pair 8-bit password-credential violation decreases; for such an example, let us assume that it is updated to 0.3 by the learning algorithm;
- The new data obtained from the source s_k will receive a lower confidentiality score sec_{conf}, which is recomputed to 0.65.

As a consequence, a user who wants to receive data from the source s_k will be aware that they have a level of confidentiality not greater than sec_{conf}, so there is a $((1 - sec_{conf}) * 100)\%$ risk of a confidentiality attack.

Knowledge and metadata required to properly assess data quality and security levels are stored in a proper format in the repository *Config*. Such a unit contains all the configuration parameters required for the correct management of the IoT system (e.g., how to calculate quality properties on the basis of the data type, which attacks or security countermeasures to consider), represented in JSON format (as described in Sect. 6). Therefore, *Analyzers* periodically query the *Config* storage unit in order to know which rules shall be used.

6 Prototype and Validation

The NOS system presented in Sect. 3 has been implemented[1] by using the following technologies: (i) the *Node.JS* platform[2] has been used for the platform implementation; (ii) *MongoDB*[3] for storage management; (iii) Mosquitto[4] for the publish/subscribe system. Modules interact among themselves via *RESTful* services.

We deployed a prototypical service middleware platform able to manage a large amount of data from heterogeneous devices with lightweight modules and interfaces working in a non-blocking manner to perform data analysis, discovery and query [8]. Such a platform is innovative for different aspects. First of all, one or more NOSs can be deployed in a distributed manner without using a peer-to-peer management, since they are completely independent from each other. This is a novel approach with respect to the conventional ad hoc centralized IoT solutions. Moreover, such solutions are often hardly reconfigurable [34], because they are conceived for very specific applications, based on a vertical silo-based approach. The middleware presented in this work supports, instead, dynamic reconfiguration and can be remotely orchestrated through Internet/intranet protocols, which are based on open standards (see Sect. 3).

As just said, a great advantage of this approach is that changes in the platform can be performed in a non-blocking manner; it is possible to introduce new modules, duplicate the existing ones or remove them without restarting the whole system. Furthermore, the use of the non-relational *MongoDB* database allows the platform to be flexible since the data model can dynamically evolve. Finally, we obtain good performance in data access, especially in read/write operations using the in-memory capability of *MongoDB*: the IoT-generated data contained in *Raw Data* and *Normal-*

---

[1]The code is released as open source under a permissive license https://bitbucket.org/alessandrarizzardi/nos.

[2]http://nodejs.org/.

[3]http://www.mongodb.org/.

[4]http://mosquitto.org.

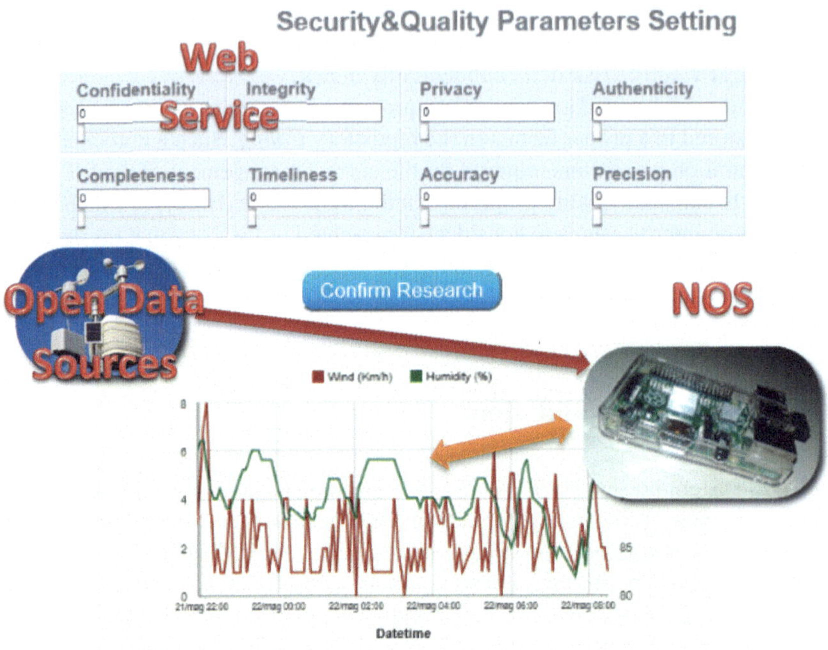

Fig. 4 User dashboard

ized Data repositories are not persistent; only the databases *Config* and *Sources* are persistent in the platform.

We tested the NOS platform by deploying it on a Raspberry Pi and connecting it to a number of open data feeds. In particular, we exploited six sensors at the meteorological station in the city of Campodenno (Trentino, Italy) that provide real-time data referred to temperature, humidity, wind speed, energy consumption and air quality. Data are transmitted by a web service that exposes them in *JSON* format, and the NOS retrieves them through HTTP *GET* requests. According to the system presented in Sect. 3, data are analyzed from a security and quality perspective following the methods presented in Sects. 4 and 5 and then transmitted to the MQTT broker.

Through a simple visualization, service users can set their preferences in terms of security, privacy and data quality and access the metadata calculated for the incoming values. The dashboard is shown in Fig. 4.

For testing the effectiveness of the proposed mechanisms, the system has been observed for a period of a week. Results have been discussed in [8]. We want only to highlight that, as specified in Sects. 4 and 5, each score is initially set to the maximum value (i.e., 1); values are then updated following changes in security and quality aspects. From the figures, it can be observed that some sources are characterized by a good level of authentication, but, at the same time, by a low level of reliability in terms of confidentiality, integrity and privacy (e.g., source 6 in Fig. 5) and viceversa

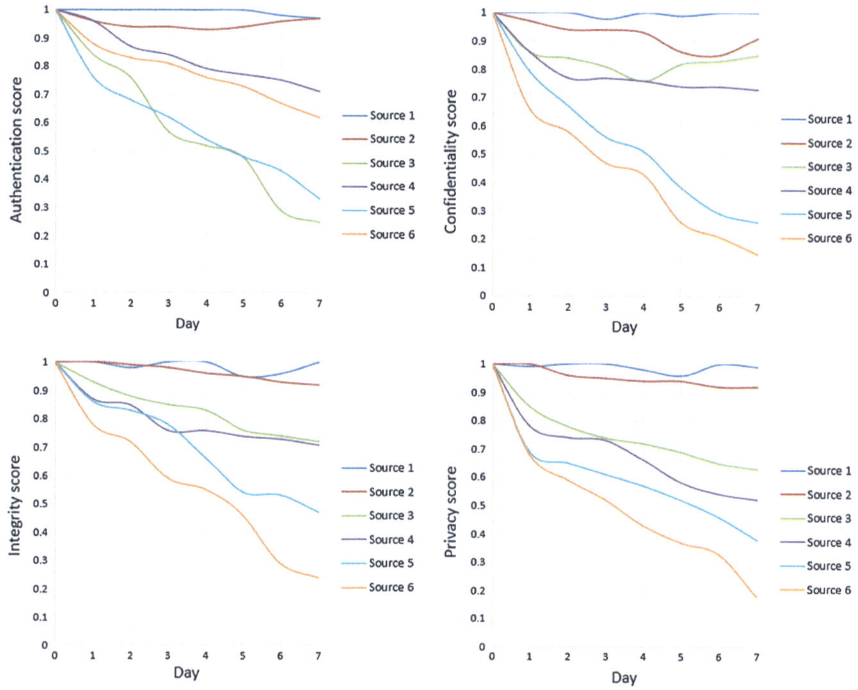

Fig. 5 Security score evaluation

(e.g., source 3 in Fig. 5). From the data quality perspective, in some sources, the provided data present good levels of completeness and timeliness, but poor accuracy and precision levels (e.g., source 4 in Fig. 6).

The experiments confirmed the usefulness of the presented approach in empowering the users to retrieve only the data that meet their requirements.

Other useful metrics about NOSs' performance, in terms of overhead, memory occupancy, computational load and latency, have been evaluated in other recent work, which include as follows: (i) the integration with an enforcement framework [35], conceived for guaranteeing a fine-grained access control against violation attempts; moreover, it allows to enforce policies specifically related to the security and data quality levels presented in this chapter; (ii) a protocol, named authenticated publish/subscribe (AUPS), for the authentication of the communications taking place via MQTT [36]; it is able to manage the disclosure of data on the basis of the policies associated to the defined topics; (iii) the integration of two key management systems, originally conceived for wireless sensor networks and adapted to the NOS platform [37]; they provide NOSs with the capabilities of handling the distribution and the replacement of the encryption keys among users and data sources. In all these cases, NOS platform demonstrated a good trade-off between a correct behavior and efficiency.

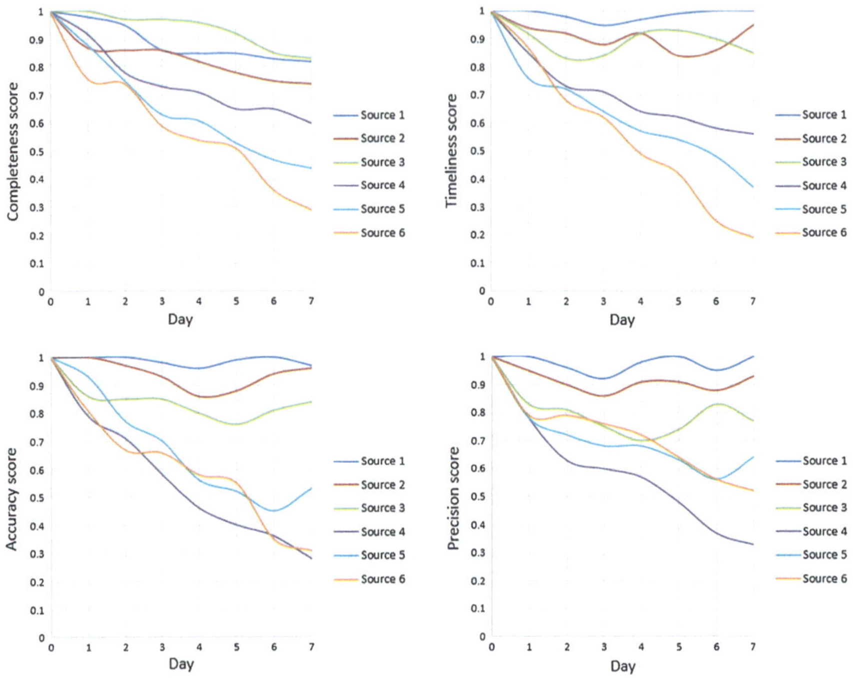

Fig. 6 Quality score evaluation

7 Conclusions

In this chapter, we have presented the design and a prototypical implementation of a distributed IoT middleware layer, named NOS, able to manage heterogeneous data sources, to provide a uniform, consistent data representation and to provide data services to manage and filter data on the basis of their related security and data quality metadata. Such an architecture improves data governance in IoT environments, since, on the one hand, it manages efficiently data and, on the other hand, it also addresses security and data quality issues by improving the user's awareness about the reliability of the accessed data. In this way, it is possible to provide data that are fit for the use in a specific context/application and thus valuable results.

The effectiveness of the proposed solution has been validated through the implementation of a real prototype of the NOS platform. Future work will focus on the design and development of new methods for assessing further data quality dimensions and for dealing with other types of sources. Moreover, new sources will be considered for the evaluation of the proposed architecture.

Acknowledgements This work is partially funded by the Department of Theoretical and Applied Sciences of University of Insubria (Varese, Italy).

References

1. D. Miorandi, S. Sicari, F. De Pellegrini, I. Chlamtac, Internet of things: vision, applications and research challenges. Ad Hoc Netw. **10**(7), 1497–1516 (2012)
2. S. Sicari, A. Rizzardi, L.A. Grieco, A. Coen-Porisini, Security, privacy and trust in internet of things: the road ahead. Comput. Netw. **76**, 146–164 (2015)
3. A. Coen Porisini, P. Colombo, S. Sicari, *Privacy aware systems: from models to patterns, igi*, global edn. (Industrial and Research Perspectives, Software Engineering for Secure Systems, 2011)
4. L. Berti-Equille, J. Borge-Holthoefer, *Veracity of Data: From Truth Discovery Computation Algorithms to Models of Misinformation Dynamics*, ser. Synthesis Lectures on Data Management. Morgan & Claypool Publishers, 2015. doi:10.2200/S00676ED1V01Y201509DTM042
5. S. Sicari, A. Rizzardi, C. Cappiello, A. Coen-Porisini, A NFP model for internet of things applications, in *Proceedings of IEEE WiMob*, Larnaca, Cyprus, Oct 2014, pp. 164–171
6. S. Sicari, C. Cappiello, F.D. Pellegrini, D. Miorandi, A. Coen-Porisini, A security-and quality-aware system architecture for Internet of Things. Inf. Syst. Front. 1–13 (2014)
7. A. Rizzardi, D. Miorandi, S. Sicari, C. Cappiello, A. Coen-Porisini, Networked smart objects: moving data processing closer to the source, in *2nd EAI International Conference on IoT as a Service*, Oct 2015
8. S. Sicari, A. Rizzardi, D. Miorandi, C. Cappiello, A. Coen-Porisini, A secure and quality-aware prototypical architecture for the internet of things. Inf. Syst. **58**, 43–55 (2016). doi:10.1016/j.is.2016.02.003
9. A. Al-Fuqaha, M. Guizani, M. Mohammadi, M. Aledhari, M. Ayyash, Internet of things: A survey on enabling technologies, protocols, and applications. IEEE Commun. Surv. Tutorials **17**(4), 2347–2376, Fourthquarter (2015)
10. IOT-A project, http://www.iot-a.eu/
11. IOT-EST project, http://ict-iotest.eu/iotest/
12. European FP7 IoT@Work project, http://iot-at-work.eu
13. EBBITS project, http://www.ebbits-project.eu/
14. Z. Yang, Y. Yue, Y. Yang, Y. Peng, X. Wang, W. Liu, Study and application on the architecture and key technologies for iot, in *2011 International Conference on Multimedia Technology*, 747–751 July 2011
15. M.A. Razzaque, M. Milojevic-Jevric, A. Palade, S. Clarke, Middleware for internet of things: a survey. IEEE Internet Things J. **3**(1), 70–95 (2016)
16. M. Palattella, N. Accettura, X. Vilajosana, T. Watteyne, L. Grieco, G. Boggia, M. Dohler, Standardized protocol stack for the Internet of (important) Things. Commun. Surv. Tutorials IEEE **15**(3), 1389–1406 (2013)
17. I. Bagci, S. Raza, T. Chung, U. Roedig, T. Voigt, Combined secure storage and communication for the Internet of Things, in *2013 IEEE International Conference on Sensing, Communications and Networking, SECON 2013*, New Orleans, LA, United States, 523–631 June 2013
18. D. Boswarthick, O. Elloumi, O. Hersent, *M2M Communications: A Systems Approach*, 1st edn. (Wiley Publishing, 2012)
19. D. Conzon, T. Bolognesi, P. Brizzi, A. Lotito, R. Tomasi, M. Spirito, The VIRTUS middleware: an XMPP based architecture for secure IoT communications, in *2012 21st International Conference on Computer Communications and Networks, ICCCN 2012*, Munich, Germany, 1–6 July 2012
20. A. Gòmez-Goiri, P. Orduna, J. Diego, D.L. de Ipina, Otsopack: lightweight semantic framework for interoperable ambient intelligence applications. Comput. Hum. Behav. **30**, 460–467 (2014)
21. C.H. Liu, B. Yang, T. Liu, Efficient naming, addressing and profile services in internet-of-things sensory environments. Ad Hoc Netw. **18**, 85–101 (2013)
22. Usable trust in the Internet of Things, http://www.utrustit.eu/
23. BUTLER project, http://www.iot-butler.eu

24. A. Karkouch, H. Mousannif, H.A. Moatassime, T. Noel, Data quality in internet of things: A state-of-the-art survey. J. Netw. Comput. Appl. **73**, 57 – 81 (2016), http://www.sciencedirect.com/science/article/pii/S1084804516301564
25. P. Barnaghi, A. Sheth, On searching the internet of things: requirements and challenges. IEEE Intell. Syst. **31**(6), 71–75 (2016)
26. B. Guo, D. Zhang, Z. Wang, Z. Yu, X. Zhou, Opportunistic IoT: exploring the harmonious interaction between human and the internet of things. J. Netw. Comput. Appl. **36**(6), 1531–1539 (2013)
27. A. Metzger, C.-H. Chi, Y. Engel, A. Marconi, Research challenges on online service quality prediction for proactive adaptation, in *Software Services and Systems Research—Results and Challenges (S-Cube), 2012 Workshop on European*, June 2012, pp. 51–57
28. Y. Qin, Q.Z. Sheng, N. J. Falkner, S. Dustdar, H. Wang, A.V. Vasilakos, When things matter: A survey on data-centric internet of things. J. Netw. Comput. Appl. **64**, 137–153 (2016), http://www.sciencedirect.com/science/article/pii/S1084804516000606
29. IBM and eurotech, mqtt v3.1 protocol specification, http://public.dhe.ibm.com/software/dw/webservices/ws-mqtt/mqtt-v3r1.html
30. C. Cappiello, F.A. Schreiber, Quality- and energy-aware data compression by aggregation in WSN data streams, in *Proceedings of the 2009 IEEE International Conference on Pervasive Computing and Communications*. (Washington, DC, USA: IEEE Computer Society, 2009), pp. 1–6
31. A. Klein, W. Lehner, Representing data quality in sensor data streaming environments. J. Data Inf. Qual. **1**(2), 10:1–10:28 (2009)
32. T. Roosta, S. Shieh, S. Sastry, Taxonomy of security attacks in sensor networks and countermeasures, in *The first IEEE International Conference on System Integration and Reliability Improvements*, vol. 25 (2006) p. 94
33. G. Tesauro, *Practical Issues in Temporal Difference Learning* (Springer, 1992)
34. I. Mashal, O. Alsaryrah, T.-Y. Chung, C.-Z. Yang, W.-H. Kuo, D.P. Agrawal, Choices for interaction with things on Internet and underlying issues. Ad Hoc Netw. **28**, 68–90 (2015)
35. S. Sicari, A. Rizzardi, D. Miorandi, C. Cappiello, A. Coen-Porisini, Security policy enforcement for networked smart objects. Comput. Netw. **108**, 133–147 (2016)
36. A. Rizzardi, S. Sicari, D. Miorandi, A. Coen-Porisini, AUPS: an open source authenticated publish/subscribe system for the internet of things. Inf. Syst. **62**, 29–41 (2016)
37. S. Sicari, A. Rizzardi, D. Miorandi, A. Coen-Porisini, Internet of Things: security in the keys, in *12th ACM International Symposium on QoS and security for wireless and mobile networks*, Malta, 129–133 2016

Advancing Cognitive Cities with the Web of Things

Sara D'Onofrio, Simone Franzelli and Edy Portmann

Abstract Currently, the Internet of Things (IoT) is employed to establish connections among smart things as well as between smart things and individuals. It is used in various applications for smart cities. However, the IoT has several drawbacks, such as a lack of common standards, which would be required if as many things as possible are to be connected. The Web of Things (WoT) (i.e., the IoT extended using Web standards) possesses common standards and has many other advantages over the IoT. This article elaborates on both approaches, compares them, and summarizes the potential uses of the WoT in cognitive cities. With the WoT, processes in cognitive cities can be simplified and living standards improved. Thus, the WoT is suitable for addressing the challenges faced by today's cities.

Keywords Cognitive city · Cognitive computing · Collective intelligence · Internet of things · Smart city · Soft computing · Web of Things

1 Introduction

The number of available data in a city is becoming so large that it is difficult to efficiently process these data (cf. big data [1]). To use such a quantity of data on behalf of urban residents and improve their user experience, cities must recognize, collect, and analyze these data [2]. A solution to this challenge is to develop information and communications technologies (ICT) that can assist in the management of urban data.

S. D'Onofrio (✉) · S. Franzelli · E. Portmann
Institute of Information Systems, University of Bern, Bern, Switzerland
e-mail: sara.donofrio@iwi.unibe.ch

S. Franzelli
e-mail: simone.franzelli@iwi.unibe.ch

E. Portmann
e-mail: edy.portmann@iwi.unibe.ch

R.R. Yager and J. Pascual Espada (eds.), *New Advances in the Internet of Things*, Studies in Computational Intelligence 715, DOI 10.1007/978-3-319-58190-3_5

Smart city refers to the use of ICT (e.g., Web-based services) in a city when accessing, processing, and using information to socially, ecologically, and efficiently develop the city and to improve living standards [3]. In this context, the Internet of Things (IoT) plays an important role because the introduction of innovative services improves the experiences of residents and thus their quality of life [4].

By embedding short-range mobile transceivers in a wide array of everyday items (e.g., mobile telephones), new forms of interaction between residents and smart things and between smart things can be established. In this manner, the IoT is created, and the ubiquity of the Internet is increased [5]. Enabling things to communicate with one another through a highly distributed network of devices enables a city to collect data in all types of situation and to share these data with its residents when they can be useful (e.g., data on traffic disturbances, weather alerts) [2, 5]. Particularly in the context of big data, it is crucial not only to be able to collect large amounts of raw data (e.g., through sensors) but also to convert the collected data into practical information for residents [6].

The IoT only represents the starting point of this development. In the future, as a result of the rapid development of the Web, a Web of Things (WoT) will probably connect residents with their cities. To date, the WoT has no clear definition. However, a small number of researchers (e.g., [7–9]) have attempted to establish one. In particular, Guinard and Trifa [10] have promoted this effort and defined the WoT as a specialization of the IoT that uses what made the Web so successful: global information access [11]. Thus far, the physical world and the Web have been separated. As an interface, a person is required, who connects the two realms by finding, integrating, and using information and services from both in a meaningful way [12]. The evolution of the Web makes it possible to connect smart things to the Web and to a large number of developers to effectively construct interactive, innovative applications that mimic reality (i.e., a blend of the physical and digital realms) [9].

The application of the WoT can influence city development (i.e., from smart to cognitive cities). The cognitive city is understood as an enhancement of the smart cities [13] by adding cognitive computing as well as cognition and learning theories, such as connectivism [14]. Connectivism describes a constant learning process with a simultaneous nurturing of connections. Not only humans but also other knowledge carriers (e.g., computer systems) can access data knowledge from other knowledge carriers. Thus, the acquisition and maintenance of knowledge is expanded from the personal dimension to multi-agent dimensions (e.g., human to human, computer system to computer system, human to computer system, computer system to human) [15]. By applying ICT, a city can use the knowledge, which has been shared with its knowledge carriers, to understand and learn from its residents. In this way, a city can recognize and react to changes in resident behavior [16]. Through the mutual communication, systems and residents learn from one another and build their common knowledge (cf. collective intelligence [17] and urban intelligence [18]). Therefore, Web standards, such as the WoT, are becoming more important because they facilitate the connection between urban systems and devices and thus optimize the interaction between residents and the city.

In this article, we compare the IoT with the WoT and demonstrate how the WoT can help smart cities evolve into cognitive cities. This article represents an outline of the current state of work-in-process and is in line with design science research [19]. The article is structured as follows: The transition from smart city to cognitive city is outlined in Sect. 2. Section 3 introduces the IoT, while Sect. 4 presents the WoT. Section 5 compares the IoT and the WoT in the context of smart and cognitive cities. Section 6 concludes the article.

2 From Smart City to Cognitive City

Although smart city is a frequently used term, a clear, consistent understanding of the concept among practitioners and scientists is lacking [2]. According to Portmann and Finger [3], a smart city is characterized by various concepts (e.g., smart democracy, smart mobility, smart work) that facilitate the interconnection of residents and modern technologies. Therefore, to enable such interconnection and the development of a city into a smart city, well-functioning ICT is required. Here, applications in many areas (e.g., health care, logistics, security) are imaginable (cf. [2]). Using such applications and thus interacting and sharing information with their city, residents become smart while further developing and shaping their city [3]. Newly developed systems (e.g., cloud-based social feedback, crowdsourcing, predictive analytics) improve the interaction between the smart resident and the city. The more input that residents have in shaping a system (i.e., a smart city), the higher their satisfaction (i.e., the better their user experience) [2]. Thus, the smart city can be understood as a sociotechnical system that aims to maintain a balance between efficiency and technology on the one hand and the happiness of its residents on the other hand [3].

A city's smartness can be described in terms of understanding, learning, and self-awareness. That is, a city can understand its own processes and learn from and reflect on them. It is important that the smart city collaborates with many disciplines (e.g., architecture, computer sciences, politics, business) and involves them in its development [2, 3]. Collectively, residents and cities can act in such a way that they behave more intelligently than they would individually, and in this manner, urban intelligence (i.e., collective intelligence [17] in a smart city) is achieved [18]. To improve the relationship between a city and its residents, it is crucial that the city understand resident needs. In this context, connectivism [14] is a significant concept. Here, not only personal experiences and perceptions but also those of others are important. Using advanced ICT to share these experiences and perceptions, the city can retrieve relevant information from its residents. Learning algorithms included in the underlying urban processes enable the city to extract patterns [20] and to understand and learn from these patterns. This learning concept is a basic principle of cognitive computing.

Cognitive computing represents an enhancement of semantic computing (which emerged from the Semantic Web [21]). Semantic computing simplifies and automates processes through definitions, models, and queries to extract the meaning of the data, while cognitive computing refers to the ability of systems to reason.

Cognitive systems can act consciously, critically, and logically to extract knowledge [22]. The application of cognitive computing and additional, new components (i.e., cognition and the related cognitive systems) support the further development of smart cities and optimize their urban processes [13]. The underlying cognitive systems can cope with natural language and perceptions and learn from the user's (i.e., the resident's) past behavior [2]. That is, these systems recognize changes in behavior and can react to them [16].

Therefore, using the advantages of cognitive computing, the city can successfully respond to challenging situations, particularly those that involve imprecise language and perceptions using cognitive systems [2, 13]. The interaction between residents and advanced ICT is optimized, which enlarges the knowledge base inside a city [3]. Because of the increased knowledge—based on the experiences and behavior of the system user (i.e., the resident)—urban processes can be performed more realistically and efficiently [16], provide more opportunities for residents to develop themselves and increase the city's attractiveness [2].

Several companies (e.g., Accenture, IBM, Microsoft) have begun applying cognitive computing in business and IT services to generate higher revenues, solve problems more efficiently, make better decisions, and improve the efficiency of applications [2]. Thus, the application of cognitive computing in urban development is of substantial interest.

3 Internet of Things

This section introduces the IoT, describes several of its technical aspects, and presents examples of IoT applications in cities.

3.1 Definition

The Internet, which is a global system of interconnected computer networks, uses the standard Internet protocol suite (TCP/IP) to link numerous devices worldwide to carry a substantial amount of data or information. The IoT functions in the same manner as the Internet (i.e., collects and shares information) by combining networks of various physical objects using network connectivity [6, 10].

The IoT's evolution can be viewed as a movement from the current Internet to a network of interconnected (smart) things. The IoT is based on interoperable ICT, which not only gathers information from the environment (i.e., through sensing) and interacts with the physical world (i.e., actuation) but also uses existing Internet standards to provide services for information transfer, analytics, and communications [23, 24].

The main characteristic of IoT is the integration of several technologies and communications solutions (e.g., identification and tracking technologies, wired and wireless sensors, actuator networks) to improve interaction and cooperation among

Fig. 1 Three visions of IoT

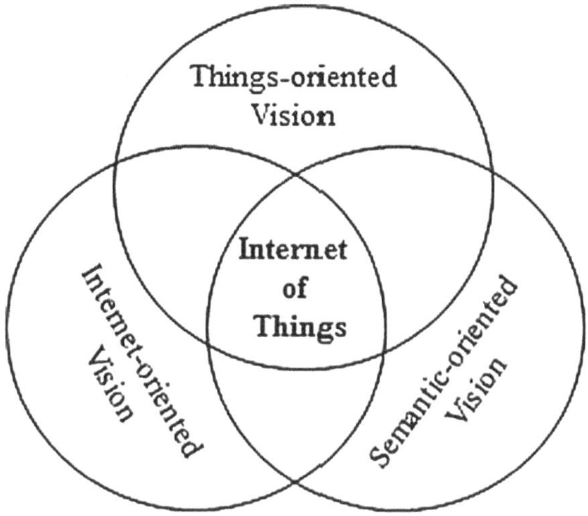

users, for example, through devices (e.g., mobile telephones, sensors), and thus to achieve common knowledge [5].

According to Atzori et al. [5], in the IoT, the following three visions converge: Internet-oriented (i.e., middleware), things-oriented (i.e., sensors), and semantic-oriented (i.e., knowledge) (Fig. 1).

The Internet-oriented vision refers to the idea of reducing the number of Internet protocols (IPs). By implementing IoT using simplified IPs, objects are made addressable and reachable from any location [5, 6]. The things-oriented vision is based on several simple items (e.g., RFID, smart items) and basic components (e.g., NFC, wireless sensors), which are applied using interfaces to connect the real world with modern ICT. The semantic-oriented vision addresses the challenge of extracting information from the increasing number of items in the Internet. In this context, the primary task is to aggregate and represent knowledge (cf. [5, 6, 25]).

From a system-level view, the IoT is a highly dynamic network system that consists of a large number of (smart) things that produce and consume information. Capable of connecting the physical realm with the digital realm and translating the data into human-readable information, the IoT opens a new era of knowledge-building [6].

3.2 Architecture

Generally, the IoT can be divided into the following 3 layers: the perception/sensing layer, the network/transmission layer, and the application layer (Fig. 2) [26, 27].

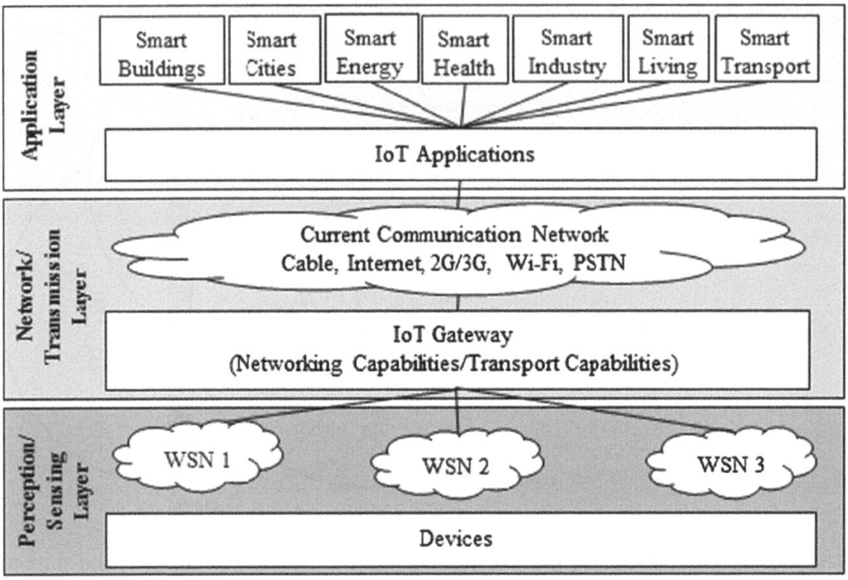

Fig. 2 IoT architecture

The primary function of the perception/sensing layer, which consists among others of two-dimensional code tags, a code reader, sensors, and sensor networks, is to perceive and identify (smart) things and to acquire and recognize information from them. The network layer (i.e., the IoT's infrastructure) is formed by all types of communication networks and the Internet. Its main parts are the IoT management center and the information center. Therefore, the network layer operates not only the network but also the information. To develop accurate application solutions, the application/transmission layer (i.e., information service system) is required that can be viewed as IoT technology combined with expertise (from industry or academia). This layer's primary task is to guarantee information sharing and information security in various areas (e.g., smart buildings, smart industry, smart transport) [26, 27]. Therefore, a typical IoT solution is characterized by many devices (i.e., smart things) that use a gateway to communicate through a network [28].

3.3 Standards and Interfaces

Standards are key for the interoperability required to improve the integration of various technologies and thus enhance the interaction between individuals and systems [29]. There are hundreds of standards from diverse organizations (e.g., IEEE,[1] NIST[2]), protocols[3] (e.g., IPv6, 6LoWPan), and IoT platforms[4] (e.g., ThingWorx, EVRYTHING, Amazon Web Services Internet of Things) to choose from [10]. Therefore, it is impossible to summarize the IoT standards.

Because there is a defined API for every standard, there are even more interfaces (API) than standards (cf. [29]). Thus, every IoT solution has its own defined API, which enables easy integration with existing applications and integration with other IoT solutions [28].

3.4 IoT Applications in Cities

Many researchers (e.g., [4, 5, 30]) have conceptualized and tested IoT applications in the urban context. A literature review, Table 1, shows examples of urban IoT applications.

Several of the studies cited in the table aim to improve urban mobility by monitoring roads and parking spaces (cf. [4, 31, 32]). Others try to decrease energy consumption and environmental pollution (cf. [5, 33, 34]). Still others collect data on buildings and infrastructure as well as general environmental data (cf. [5, 35]). There is also an application from the shopping domain (cf. [36]). In sum, these examples of urban applications indicate that the IoT has the potential to improve city development through the integration of human and machine.

4 Web of Things

This section introduces the WoT, describes several of its technical aspects, and presents examples of WoT applications used in cities.

4.1 Definition

The Web is constantly evolving (i.e., from Web 1.0 to Web 4.0 [37]). Thus, today, anyone can access Web servers from personal devices (e.g., computer, mobile telephone), and services are increasingly provided on the Internet using Web applications [8].

---

[1]cf. https://www.ieee.org/index.html.

[2]cf. https://www.nist.gov/.

[3]cf. http://www.postscapes.com/internet-of-things-protocols/.

[4]cf. http://internetofthingswiki.com/top-10-iot-platforms/634/.

Table 1 IoT applications in cities

Source	Approach	Description
[5]	City information model	The city continuously monitors the status and performance of its buildings and infrastructure (e.g., cycle paths, rail lines) and shares the information with third parties through APIs
[35]	Low carbon open data network	Using low-power, low-cost sensing equipment, environmental data are collected in real time. Open access to these data is provided to residents via online services, which means that they can develop applications based on the data
[31]	RDF stream processing	A travel planner application, which recommends the best route based on live data from traffic sensors while considering factors such as the user's transport mode, traffic congestion, and estimated arrival time
[32]	Road condition application	An application that provides road condition alerts based on data from embedded sensors in the smartphones of vehicle users
[36]	Smart shopping	A smart shopping environment that tells merchants when to inform citizens regarding shopping offers based on the analysis of city-context data (e.g., city agenda, parking data)
[33]	Energy management	The application draws data from different sources (e.g., heat and electricity meters) to improve the use of energy in commercial and residential areas
[4]	Traffic control systems	IoT technologies are applied to monitor traffic and parking spaces in cities as well as to offer traffic routing advice
[34]	Semantic framework	A framework that collects data and models them for specific IoT applications (e.g., detection of vehicle pollution) that are based on semantic and machine-learning technologies

The WoT is a means of accessing surrounding devices through Web applications [8]. The WoT's underlying idea is for each (smart) thing to have its own Web page and thus be available for indexing by search engines. Subsequently, one can search for the thing and access it directly from a Web browser [38]. By reusing Web standards and adapting technologies and patterns commonly used for traditional Web content, objects (i.e., things) of daily life can be connected and fully integrated into the Web [5, 39]. Therefore, the WoT is an environment in which everyday objects (e.g., buildings, traffic lights, commodities) are identifiable, recognizable, and controllable through the Internet using Semantic Web standards. Thus, a large number of things appear on the Web, which means that seamless communication between individuals and (smart) things is provided. Similar to the IoT, the WoT seeks to bridge the gap between the physical and digital realms using a common platform (i.e., the Web) that is accessible to everyone [9].

According to Trifa [40], the WoT is based on five pillars of the modern Web architecture (Fig. 3).

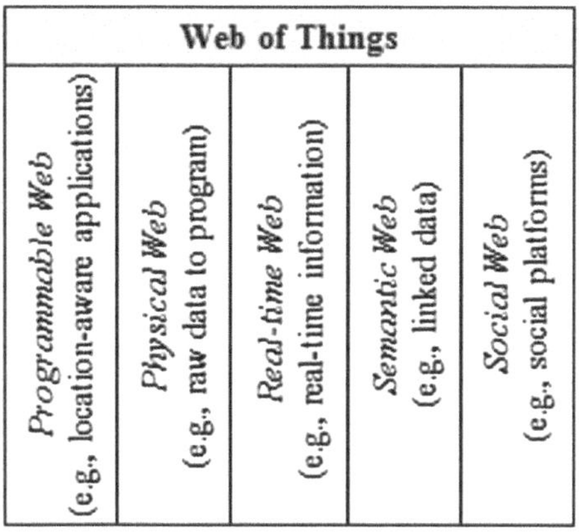

Fig. 3 WoT characteristics

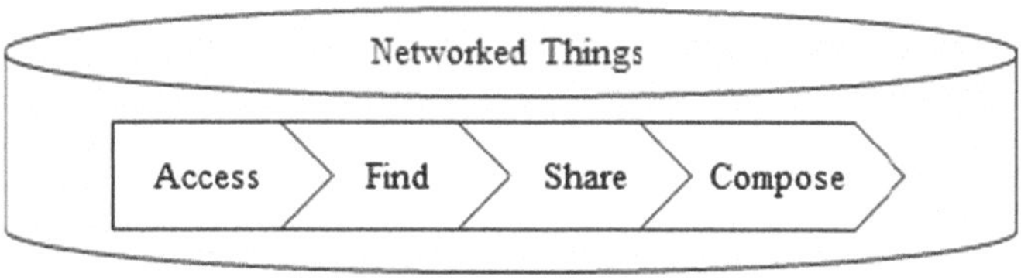

Fig. 4 WoT architecture

4.2 Architecture

The WoT has the following four layers: *Access*, *Find*, *Share*, and *Compose* (Fig. 4). These layers more thoroughly integrate (smart) things into the Web and make them more accessible for applications and individuals [10].

In *Access*, the primary task consists of transforming a (smart) thing into a programmable Web thing with which other devices can easily communicate. The layer *Find* ensures that a device is findable and automatically usable by other WoT applications. *Share* specifies how the data generated by (smart) things can be efficiently shared over the Web. Finally, *Compose* creates in a simple way applications that involve (smart) things and virtual Web services [10]. Thus, the aim is not to apply the Web as a transport infrastructure but to make devices an integral part of the Web [39].

4.3 Standards and Interfaces

The WoT does not create completely new standards. Instead, it reuses the well-known Web standards [39] used in the physical (e.g., Beacon), programmable (e.g., REST, HTTP), semantic (e.g., RDF, OWL), real-time (e.g., WebSockets), and social (e.g., profile standards) Webs [40].

Web standards ensure that data can be rapidly and easily moved across systems. HTTP and REST (cf. [41]) are the most frequently recommended Web services for offering public access to data available on the Web. With REST constraints, the interaction between the components (because each component of the system complies with those constraints) is well defined and thus predictable. Therefore, (smart) things can be smoothly integrated into the Web by making them findable for Web users through a RESTful interface (API) using HTTP [10].

4.4 WoT Applications in Cities

Although the WoT remains an emerging approach, a small number of researchers (e.g., [42–44]) have theoretically applied it in a city context. A literature review, Table 2, shows examples of WoT applications used in cities.

Table 2 WoT applications in cities

Source	Approach	Description
[42]	Traffic SenseBox	A sensor platform focused on easy accessibility via the Web, with a built-in ultrasonic sensor that can determine traffic density by counting the number of passing cars
[7]	WebIoT	A Web application framework that aims to improve the interaction among things and between things and humans
[45]	Intelligent vehicle system	Road users can apply a traffic information service, which shows road obstacles and congestion levels and is personalized through WoT-API for their requirements
[39]	Energie visible	This project provides a Web dashboard, which enables individuals to visualize and better understand, monitor, and control the energy consumption of household appliances
[43]	Smart home	Using the Web's infrastructure, Web applications can be designed that benefit a large number of simultaneous users, who can fully automate their houses (i.e., the smart home)
[30]	Landsliding early warning	Sensor nodes that are implemented in a landsliding high-risk zone collect and display real-time telemetry observations of, e.g., soil movement and precipitation
[51]	Multimedia remote controller	A prototype based on a Web application that can exchange information with nearby electronic devices and on a mobile Bluetooth application
[44]	Smart farm	A prototype relying on Semantic Web standards that uses livestock monitoring technologies, environmental sensors, and an ontology-enabled architecture for personal real-time alerts for on-farm situation awareness

Currently, it is difficult to find WoT applications in actual use in cities. Several of the applications mentioned in Table 2 are envisaged use cases and have not yet been applied in real-world scenarios. There are various areas in which such WoT applications may be used, for example, in the measurement and visualization of traffic flows (cf. [42, 45]) or energy consumption (cf. [39]). WoT applications could also be helpful in the development and functioning of smart homes (cf. [43]), smart farms (cf. [44]), or natural disaster warning systems (cf. [30]). More generally, a Web application framework aimed at improving the interaction among things and between things and humans [7] as well as a system that facilitates the sharing and control of the access to resources in the WoT [46] are presented in the table.

As in the case of the IoT (because the WoT is specialization of the IoT [10]), one can conclude that the WoT has the potential to improve city development through a closer interaction and integration of residents and machines.

5 The Internet of Things Compared with the Web of Things

First, a general comparison of the IoT and the WoT is made, followed by a comparison of the main aspects of both approaches in the context of smart and cognitive cities.

5.1 General Comparison

The IoT only involves the interconnection of (smart) things with the Internet, without considering any technology or network structures [8]. Therefore, while the Internet is the correct option to connect physical things, the option for a universal platform on which to construct applications using things is the Web [9]. To illustrate this point, Table 3 shows a literature-based comparison of both approaches, in which '+' indicates positive (incl. potential) impacts and '–' indicates negative impacts (incl. threats). In this comparison, only decisive criteria are included (i.e., those that are mentioned most frequently in the consulted literature). Additional criteria (e.g., automation, quality of life, sustainability) were considered. However, because the IoT and the WoT have similar advantages in these fields, they have been omitted. Therefore, Table 3 only includes criteria with respect to which major differences appear between the IoT and the WoT and therefore is incomplete.

As Table 3 shows, the WoT has several advantages over the IoT. For example, the WoT is much easier to maintain and program. It is also substantially more secure than the IoT, and its standards are more universally applicable than the fragmented IoT standards. Only in the aspect of privacy do both systems display problems. However, even in this respect, the WoT is one step ahead of the IoT.

Table 3 Comparison of the IoT and the WoT

	IoT	WoT
Maintainability	• Significant effort required to write custom convertors for each new device [10] • Long-term maintainability vulnerabilities (because of competitive cost and technical constraints) [47] • Improved maintainability of the systems through integrators [50]	• Existing widespread (Web) technologies enable substantial ease of development [8] • Maintaining Web applications is more cost-efficient [51] • No risk that the Web will suddenly cease to function and require an upgrade [10]
Privacy	• Potential harm is amplified in the IoT by the scale and greater intimacy of personal data collection [47] • Privacy breach (i.e., when a thing is put online, it remains online) [53] • Privacy requirement in the IoT is currently only partially covered [52]	• Potential privacy violations (i.e., Web services having drawbacks) [46] • Public sharing might result in serious privacy implications [39] • Standard protocols for securely encrypting data between clients and servers on the Web [10]
Programmability	• Complex ease of design and development [48] • Much computing power for identification/addressing schemes [49] • High barrier for adoption (i.e., complex IoT protocols) [10]	• Easier; surrounding devices are accessed through Web applications [8] • Open ecosystem (i.e., creating applications using standard Web services) [46] • Same programming model for interaction with Web API [10]
Security	• Vulnerable to attack (e.g., unattended components, wireless communications, low capabilities of energy and computing resources) [5] • Possibility of personal data being stolen [53] • Security problems [27]	• Secure interactions with HTTPS [8] • Less risky (i.e., constantly tested, updated, and fixed systems) [10] • Authenticated and secure communication between clients and gateways with HTTPS and OAuth [40]
Standards	• Complex standards landscape [29] • Standards funded and governed by corporations are not neutral [10] • Risk of fragmentation and lack of adoption of adequate standards [4]	• Adoption of IP (i.e., open standards for thing communications) [7] • Promising results when using Web standards (i.e., easily accessible) [39] • Open and free standards [10]

The next section treats most of these aspects in more detail. However, it should be kept in mind that this comparison is only a literature-based one and thus does not represent an exhaustive list.

5.2 Benefits for Smart and Cognitive Cities

In 2011, the number of interconnected devices exceeded the world's human population. Currently, there are approximately 9 billion interconnected devices. The number predicted for 2020 is approximately 24 billion devices [23]. More than ever, the IoT and the WoT concepts are becoming interesting for cities.

The aim of a smart city is to more efficiently use resources to improve the quality of services and thus to increase the city's attractiveness [2]. Embedding devices into everyday physical objects and making them smart enable the integration of such objects into the global cyberphysical infrastructure [4, 6]. Thus, the interconnection of the physical and virtual realms can improve the management of cities and urban processes in various fields (e.g., education, health, logistics) [2]. Using IoT, a smart city confronts certain obstacles (e.g., long-term maintainability vulnerabilities [47], complexity in programming [48]). More precisely, the city must invest a significant effort to program and maintain custom convertors, identification, and addressing schemes for each device [10, 49, 50].

Such programming is coupled with standards. Because there is a large, complex standards landscape that is to a certain extent funded and governed by corporations, IoT development is complicated [29]. Therefore, the use of Web standards in the context of cognitive cities is particularly interesting. These open, free standards make it substantially easier for cities to share data with their residents [2, 10]. The possibility of accessing surrounding devices through Web applications and the use of open standards make these devices easily accessible in a universal way [7, 8, 39].

This open ecosystem of digitally augmented smart things makes it easier for developers to process real-time data from various fields (e.g., traffic pollution, public transportation) [46]. Because there is no risk that the Web will suddenly cease to function, the maintenance of such Web applications is cost-efficient, which enables substantial ease of development [8, 10, 51]. The sharing of real-time information between smart things and city residents results in a learning cycle and pattern recognition. Through this mutual learning process (i.e., connectivism [14]), collective intelligence [17] (i.e., urban intelligence [18]) can be achieved.

Regarding the maintainability, programming, and standardization of applications, it seems that the Web offers more potential for urban development than the Internet does. However, both approaches are confronted by issues of privacy and security. In the IoT, privacy requirements are generally only partially addressed [52], which makes the connected devices highly vulnerable to attack. In the worst case, personal data might be stolen [5, 53]. In the WoT, the Web continues to display several drawbacks that could have serious privacy implications [39, 46]. However, by applying the HTTP programming model, particularly HTTPS, it is possible to offer authenticated, secure communication between mobile clients and gateways [8, 40]. In addition, there is less risk of attack because Web services are constantly used, tested, updated, and fixed [10]. Even if the issues of security and privacy are difficult, the Web is better able to counter these challenges than the Internet.

6 Conclusions and Outlook

Because of the potential of connecting smart things with individuals and systems through the Internet, several studies (e.g., [54, 55]) suggest that the IoT will evolve into a future IoT (i.e., an advanced and extended version of the IoT). However, we believe that the next phase of the IoT will be the WoT (i.e., the IoT extended by the Web), particularly in the context of city development. Regarding the advantages of the WoT over the IoT maintainability, programming, and standardization, the authors recommend considering the WoT for the development of smart and cognitive cities. In particular, the following aspects make the WoT useful for cognitive cities: first, common standards and ease of programming, which enable more individuals to participate in the development of new applications; second, the higher security of communication; third, the reduced effort required to maintain the Web. It must be noted that the comparison of the two approaches is based on information sciences research and thus provides insight from only one perspective. To clearly state which approach is better (and to what extent) for city development, other perspectives (e.g., mathematical, engineering) should also be considered.

An important issue in the context of the WoT that has not been mentioned is the search process, which enables connected objects to be discovered and filtered for a given use and to be combined or associated within Web applications. Objects can only be found if they are described with identifiable and traceable information. To use these connected objects in Web applications, dedicated description languages (e.g., WSDL, WADL) are required. With the emergence of Web-enabled objects, efficient techniques that facilitate the search for connected objects will be crucial for the WoT's success [56].

In addition, the privacy and security issues should be elaborated in more detail. Even if the HTTP programming model facilitates secure communication, the Web possesses other drawbacks that cannot be ignored.

Moreover, research on how the WoT can benefit smart and cognitive cities in specific real-life cases should be expanded. Only a small number of applications of and use cases for the WoT in cities have been developed, which indicates a substantial opportunity for future research in this area (e.g., the creation of a *Google* for cities).

Acknowledgements The authors would like to thank the Institute of Information Systems at the University of Bern, particularly Anne Gerster and Michelle Ballmer, for the helpful input. We are also grateful to Joël Vogt from the WoT company EVRYTHNG,[5] who provided constant support, help, and advice regarding the IoT.

[5]cf. https://evrythng.com.

References

1. Y. Chen, E. Argentinis, G. Weber, IBM Watson: how cognitive computing can be applied to big data challenges in life sciences research. Clin. Ther. **38**(4), 688–701 (2016)
2. S. D'Onofrio, E. Portmann, Cognitive computing in smart cities. Inf. Spektrum, 1–12 (2016)
3. E. Portmann, M. Finger, Smart Cities! Ein Überblick. HMD Praxis der Wirtschaftsinformatik **52**(4), 470–481 (2015)
4. D. Miorandi, S. Sicari, F. De Pellegrini, I. Chlamtac, Internet of things: vision, applications and research challenges. Ad Hoc Netw. **10**(7), 1497–1516 (2012)
5. L. Atzori, A. Iera, G. Morabito, The internet of things: a survey. Comput. Netw. **54**(15), 2787–2805 (2010)
6. G. Misra, V. Kumar, A. Agarwal, K. Agarwal, Internet of Things (IoT)—A Technological Analysis and Survey on Vision, Concepts, Challenges, Innovation, Directions, Technologies, and Applications (An Upcoming or Future Generation Computer Communication System Technology). Am. J. Electric. Electron. Eng. **4**(1), 23–32 (2016)
7. A.P. Castellani, M. Dissegna, N. Bui, M. Zorzi, WebIoT: a web application framework for the internet of things, in *Workshop on Wireless Communications and Networking Conference Workshops (WCNCW)*, (IEEE, 2012), pp. 202–207
8. S. Duquennoy, G. Grimaud, J.J. Vandewalle, The web of things: interconnecting devices with high usability and performance, in *International Conference on Embedded Software and Systems*, (2009), pp. 323–330
9. S.S. Mathew, Y. Atif, Q.Z. Sheng, Z. Maamar, Web of things: description, discovery and integration, in *IEEE International Conferences on Internet of Things, and Cyber, Physical and Social Computing*, vol. 9, no. 15. (2011), pp. 19–22
10. D. Guinard, V. Trifa, *Building the Web of Things* (Manning, Shelter Island, 2016)
11. E. Portmann, The FORA framework—a fuzzy grassroots ontology for online reputation management. Dissertation 2012
12. D. Pfisterer, K. Römer, D. Bimschas, H. Hasemann, M. Hauswirth, M. Karnstedt, O. Kleine, A. Kröller, M. Leggieri, R. Mietz, M. Pagel, A. Passant, R. Richardson, C. Truong, SPITFIRE: Towards a Semantic Web of Things. IEEE Commun. Mag. **49**(11), 40–48 (2011)
13. E. Portmann, M. Finger, *Towards Cognitive Cities—Advances in Cognitive Computing and its Application to the Governance of Large Urban Systems*, vol. 63. (Springer International Publishing 2016)
14. G. Siemens, Connectivism: a learning theory for the digital age. Int. J. Instr. Technol. Distance Learn. **2**(1), 3–10 (2005)
15. P. Kaltenrieder, E. Portmann, S. D'Onofrio, Enhancing multidirectional communication for cognitive cities, in *2nd International Conference on eDemocracy & eGovernment*, Quito, Ecuador, (2015)
16. J.S. Hurwitz, M. Kaufman, A. Bowles, *Cognitive Computing and Big Data Analytics* (John Wiley and Sons Inc, Hoboken, New Jersey, 2015)
17. T.W. Malone, M.S. Bernstein, *Handbook of collective intelligence* (The MIT Press, Cambridge, 2015)
18. R. Moyser, S. Uffer, From smart to cognitive: a roadmap for the adoption of technology in cities, in *Towards Cognitive Cities: Advances in Cognitive Computing and Its Applications to The Governance of Large Urban Systems*, ed. by E. Portmann, M. Finger (Springer International Publishing, Heidelberg, 2016), pp. 13–35
19. A.R. Hevner, S.T. March, J. Park, S. Ram, Design science in information systems research. MIS Q. **28**(1), 75–105 (2004)
20. G. Papakostas, E. Papageorgiou, V. Kaburlasos, Linguistic fuzzy cognitive maps (LFCM) for pattern recognition, in *IEEE International Conference on Fuzzy Systems (FUZZIEEE 2015)*, Istanbul, Turkey, (2015), pp. 1–7
21. T. Berners-Lee, J. Hendler, O. Lassila, The semantic web. Sci. Am. **284**(5), 28–37 (2001)

22. S. D'Onofrio, E. Portmann, Von fuzzy-sets zu computing-with-words. Inf. Spektrum **38**(6), 543–549 (2015)
23. J. Gubbi, R. Buyya, S. Marusic, M. Palaniswami, Internet of things (IoT): a vision, architectural elements, and future directions. Future Gener. Comput. Syst. **29**, 1645–1660 (2013)
24. International Telecommunication Union, *Overview of the internet of things* (ITU, Geneva, 2012)
25. E. Portmann, P. Kaltenrieder, W. Pedrycz, Knowledge representation through graphs, in *International Conference on Soft Computing and Software Engineering*, Berkeley, California (2015)
26. M. Yun, B. Yuxin, Research on the architecture and key technology of internet of things (IoT) applied on smart grid, in *Proceedings of the International Conference on Advances in Energy Engineering*, (2010), pp. 69–72
27. H. Zhang, L. Zhu, Internet of things: key technology, architecture and challenging problems, in *IEEE International Conference on Computer Science and Automation Engineering (CSAE)*, vol. 4, (2011), pp. 507–512
28. Eclipse IoT White Paper, *The Three Software Stacks Required for IoT Architectures*, (2016), pp. 1–16
29. P.F. Drucker, Internet of things—position paper on standardization for IoT technologies, in *European Research Cluster on Internet of Things*, (2015), pp. 1–142
30. M.O. Kebaili, K. Foughali, K. FathAllah, A. Frihida, T. Ezzeddine, C. Claramunt, Landsliding early warning prototype using MongoDB and web of things technologies. Proc. Comput. Sci. **98**, 578–583 (2016)
31. F. Gao, M.I. Ali, A. Mileo, RDF stream processing for smart city applications, in *RDF Stream Processing Workshop (ESWC2015)*, (Slovenia, 2015), pp. 1–3
32. A. Ghose, P. Biswas, C. Bhaumik, M. Sharma, A. Pal, A. Jha, Road condition monitoring and alert application, in *IEEE Pervasive Computing and Communication (PerCom 2012)*, (Lugano, Switzerland, 2012), pp. 489–491
33. D. Kyriazis, T. Varvarigou, A. Rossi, D. White, J. Cooper, Sustainable smart city IoT applications: heat and electricity management & Eco-conscious cruise control for public transportation, in *International Symposium and Workshops on a World of Wireless, Mobile and Multimedia Networks (WoWMom 2013)*, (2013), pp. 1–5
34. N. Zhang, H. Chen, X. Chen, J. Chen, Semantic framework of internet of things for smart cities: case studies. Sensors **16**(9), 1–13 (2016)
35. D. Carter, Urban regeneration, digital development strategies and the knowledge economy: manchester case study. J. Knowl. Econ. **4**(2), 69–189 (2013)
36. IoT Open Platforms, BUTLER Smart Parking Trial, http://open-platforms.eu/app_deployment/butler-smart-parking-trial/. Accessed 12 Jan 2017
37. S. Aghaei, M.A. Nematbakhsh, H.K. Farsani, Evolution of the world wide web: from web 1.0 to web 4.0. Int. J. Web Semant. Technol. (IJWest) **3**(1):1–10 (2012)
38. D. Guinard, V. Trifa, Towards the web of things: web mashups for embedded devices, in *International World Wide Web Conference, Workshop on Mashups, Enterprise Mashups and Lightweight Composition on the Web (MEM 2009)*, (2009), p. 15
39. D. Guinard, V. Trifa, F. Mattern, E. Wilde, From the internet of things to the web of things: resource-oriented architecture and best practices, chapter, in *Architecting the Internet of Things*, ed. by D. Uckelmann, M. Harrison, F. Michahelles (Springer, Berlin Heidelberg, 2011), pp. 97–129
40. V. Trifa, Building blocks for a participatory web of things: devices, infrastructures, and programming frameworks. Dissertation 2011
41. S. Tilkov, REST und HTTP Einsatz der Architektur des Web für Integrationsszenarien, dpunkt.verlag (2011)
42. A. Bröring, A. Remke, D. Lasnia, SenseBox—a generic sensor platform for the web of things, in 8th *Annual International Conference on Mobile and Ubiquitous Systems: Computing,*

Networking and Services (MobiQuitous 2011), vol. 104, no. 5 (Springer Berlin, 2012), pp. 186–196
43. A. Kamilaris, A. Pitsillides, The smart home meets the web of things. Int. J. Ad Hoc Ubiquitous Comput. **7**(3), 145–154 (2011)
44. K. Taylor, C. Griffith, L. Lefort, R. Gaire, M. Compton, T. Wark, D. Lamb, G. Falzon, M. Trotter, Farming the web of things. IEEE Intell. Syst. **28**(6), 12–19 (2013)
45. T.S. Dillon, H. Zhuge, C. Wu, J. Singh, E. Chang, Web-of-things framework for cyber-physical systems. Concurr. Comput. Pract. Exp. **23**(9), 905–923 (2011)
46. D. Guinard, V. Trifa, E. Wilde, A resource oriented architecture for the web of things, in *Internet of Things (IOT)*, (IEEE, 2010), pp. 1–8
47. Internet Society Whitepaper, The internet of things: an overview—understanding the issues and challenges of a more connected world, 1–75 (2015)
48. J. Chase, *The evolution of the internet of things* (Strategic marketing, Texas Instruments, Texas Instruments Incorporated, Dallas, 2013), pp. 1–6
49. E. Fleisch, What is the internet of things? An economic perspective, economics, management, and financial market **2**, 125–157 (2010)
50. F. Jammes, *Internet of Things in Energy Efficiency, Ubiquity Symposium*, (ACM, 2016), p. 2
51. J. Pascual-Espada, V.G. Diaz, R.G. Crespo, O.S. Martinez, B.C.P. G-Bustelo, J.M. Cueva Lovelle, Using extended web technologies to develop Bluetooth multi-platform mobile applications or interact with smart things. Inf. Fusion, **21**:30–41 (2015)
52. S. Sicari, A. Rizzardi, L.A. Grieco, A. Coen-Porisini, Security, privacy and trust in internet of things: the road ahead. Comput. Netw. **76**, 146–164 (2015)
53. Z. Jun, Internet of Things, Disruption or Destruction? Amsterdam Business School (2014)
54. R. Khan, S.U. Khan, R. Zaheer, S. Khan, Future internet: the internet of things architecture, possible applications and key challenges, in *IEEE 10th International Conference on Frontiers of Information Technology (FIT)*, (2012), pp. 257–260
55. J.A. Stankovic, Research directions for the internet of things. IEEE Internet Things J **1**(1), 3–9 (2014)
56. C. Benoit, V. Verdot, V. Toubiana, Searching the 'Web of Things', in *Fifth IEEE International Conference on Semantic Computing (ICSC)* (2011), pp. 308–315

Transferring Wireless High Update Rate Supermedia Streams Over IoT

George Kokkonis, Kostas E. Psannis, Manos Roumeliotis, Yutaka Ishibashi, Byung-Gyu Kim and Anthony G. Constantinides

Abstract This paper deals with the wireless transfer of real-time high update rate supermedia data over the Internet of Things. It presents the related work on supermedia data transferring and QoE requirements. It proposes a high-level architectural design for the transport of wireless multiple supermedia streams over IoT. The most known compression techniques and flow controls for wireless sensory data transferring are analyzed. Based on these compression techniques, a new network adaptive flow control algorithm is proposed. Measurements for multihop wireless transferring of high update rate supermedia packets over IoT are presented.

Keywords IoT · Wireless communications · Supermedia · Haptics · Transport protocols · Real-time protocols · Flow control · Congestion control

---

G. Kokkonis (✉) · K.E. Psannis · M. Roumeliotis
Department of Applied Informatics, University of Macedonia, 156 Egnatia Street, 54006 Thessaloniki, Greece
e-mail: gkokkonis@uom.gr

K.E. Psannis
e-mail: kpsannis@uom.gr

M. Roumeliotis
e-mail: manos@uom.gr

Y. Ishibashi
Department of Computer Science, Graduate School of Engineering, Nagoya Institute of Technology, Nagoya 466-8555, Japan
e-mail: ishibasi@nitech.ac.jp

B.-G. Kim
Department of IT Engineering, Sookmyung Women's University Seoul Republic of Korea, 100 Kalsan-ri Tangjeong-myeon, A-san City, Chungnam 305-350, Korea
e-mail: bg.kim@sm.ac.kr

A.G. Constantinides
Department of Electrical and Electronic Engineering, Imperial College London, London SW7 2AZ, UK
e-mail: a.constantinides@imperial.ac.uk

R.R. Yager and J. Pascual Espada (eds.), *New Advances in the Internet of Things*, Studies in Computational Intelligence 715, DOI 10.1007/978-3-319-58190-3_6

1 Introduction

Over the last years, the Internet has evolved from the WWW, the simple transfer of text and static pages to the Web 2, the M2M communication, and the transfer of real-time supermedia. The new stage of Internet evolution is the Internet of Things. Internet of Things semantically means a worldwide network of interconnected object uniquely addressable, based on standard communication protocols [1]. The IoT improves efficiency and effectiveness and offers new business opportunities. Apart from person-to-person communication, the person-to-object and object-to-object communications are now evolving. Apart from humans, sensors, actuators, and smart objects are now exchanging information. IoT devices have exceeded 30 billion. The number of the message transactions is increasing rapidly. One of the challenges that the IoT has to overcome is to ensure the Quality of Service (QoS) for all participants and maximize the Quality of Experience for all users. New protocols such as the MQTT, the XMPP, the COAP, and the 6LOWPAN have already been proposed [2]. These protocols take into consideration issues such as the energy consumption, safety, and scalability. The main objective of IoT should be the Quality of Experience of the user. As Internet network conditions are constantly changing, transport protocols and flow control algorithms should monitor network conditions and modify the supermedia streams of data.

In this massive network, wireless communications are invited to play a major role. Most of the end users of IoT transfer their data wirelessly. This gives them the ability to move freely with no wire constraints. All wireless communication standards that could be used for this transmission are shown in Table 1.

The 3G/4G and the 802.11 g are the most prevailing standards for the IoT because of the available range they offer. The 3G/4G uses the cellular network to offer unlimited range, while the range of a WiFi network may be extended to only some several hundred of meters but is much faster than the 4G. The most promising standard is the 5G (fifth generation mobile networks or fifth generation wireless systems). It is supposed to be available by 2020, and its three main criteria are to offer unlimited range within its cellular network with speeds more than 1 gigabit per second, latency lower than one millisecond, and more energy efficient than 4G.

Table 1 Wireless communication standards

	Range	Data rate	Power	Frequency
ZigBee	10–75 m	20/40/250 kbps	30 mW	868/915/2400 MHz
Bluetooth	10–100 m	1–3 Mbps	2.5–100 mW	2.4 GHz
IrDA	1 m	16 Mbps	10 μW	Infrared
MICS	2 m	0.5 Mbps	25 μW	402–405 MHz
802.11 g	0.2–1 km	54 Mbps	0.1–1 W	2.4 GHz
3G/4G	Cellular based	5/12 Mbps	32–200 mW	900/1800/2300 MHz
5G	Cellular based	>1 Gbps	Not defined yet	Not defined yet

One major part of the data that will be transferred over the IoT is regarding to sensors and actuators. An important segment of this data refers to the haptic human sense. Haptics is the science of manual sensing and manipulation of objects through the sense of touch. Haptic data play a major role to the Quality of Experience of the user, transform virtual reality to augment reality, and evolve multimedia to supermedia. The main obstacle that impedes tele-haptics from flourishing is the delay and the jitter that is being encountered in the wireless IoT. Until 5G is available to public, flow control algorithms should be defined in order to mitigate congestion, latency, and packet loss in IoT under the available wireless communications.

The rest of the paper is organized as follows: Sect. 2 presents QoS requirements for haptic data transferring. Section 3 describes the proposed flow control algorithm for haptic data over the wireless IoT. Section 4 presents measurements for transferring wireless high update rate supermedia streams. Section 5 summarizes and concludes this paper.

2 Supermedia System Transferring High-Level Architectural and QoS Requirements

Real-time supermedia applications require timely delivery of video, audio, graphics, haptics, and other sensory data. In order to infer the requirements for establishing the Quality of Service (QoS) for supermedia applications, the mean opinion score (MOS) is used as a basic metric. MOS is merging the gap between the QoS and the Quality of Experience (QoE). MOS is evaluating the quality of experience that a user receives when he/she uses a supermedia service. It is observed that the Quality of Experience deteriorates mainly when the network delay is relatively high and unstable. The packet delay deviation, often called jitter, is the main reason for the instability of supermedia systems. High values of jitter and delay often occur in best effort, heavily congested networks. Such a network is the IoT.

The network conditions that have to be fulfilled in order to maximize the QoE for supermedia transferring through the IoT are depicted in Table 2 [3].

Table 2 QoS requirements for supermedia applications [3]

QOS	Haptics	Video	Audio	Graphics
Jitter (ms)	≤ 2	≤ 30	≤ 30	≤ 30
Delay (ms)	≤ 50	≤ 400	≤ 150	$\leq$ 100–300
Packet loss (%)	≤ 10	≤ 1	≤ 1	≤ 10
Update rate (Hz)	≥ 1000	≥ 30	≥ 50	≥ 30
Packet size (bytes)	64–128	$\leq$ MTU	160–320	192–5000
Throughput (kbps)	512–1024	2500–40000	64–128	45–1200

From Table 2 it is understood that haptic applications are more sensitive to jitter and delay than other common multimedia applications such as audio, video, and graphics. The delay in haptic applications has to be smaller than 50 ms while video, audio, and graphic applications must have network delay smaller than 400, 150, and 100 ms, respectively.

Especially jitter has to be so small, less than 2 ms, that haptic applications often result in system failure due to instability and lack of synchronization. One compensation technique that partially overcomes this barrier is a buffer that is being put at the receiver's side in order to deliver the packets to the receiver with a more constant rate. The drawback of this technique is that it raises the mean end-to-end delay.

The update rate in this kind of applications should be relatively high. In order to have the maximum sensation of "telepresence," the update rate should be 1000 packets per second. This high update rate is needed especially in situations where users/objects come in conduct with each other. The harder the objects are, the higher the update rate should be, so that unwanted intrusions and oscillations can be avoided. Although, interesting studies have been made for the reduction of this high update rate, it is still considered to be significantly high. The reduction of the sending rate is mainly based on Weber's Law of Just Noticeable Differences (JND) which describes the haptic human perception, and is often called the "deadband control" approach [4].

On the other hand, the throughput in this kind of applications is not very demanding. The small throughput is based on the fact that the packet size of control commands and haptic feedback is relative small. It is often smaller than 64–128 bytes. This means that for an update rate of 1000 packets per second, the throughput is 512–1024 kbps.

Furthermore, the data loss in haptic applications could be quite high, up to 10%. The high update rate often compensates for missing packets, and the packet loss event is not understood by users. Depending on the transport and the application layer protocol, most of the "normal updates" packets are sent unreliably, except of a few "crucial data" that are sent reliably [5]. The high percentage of the acceptable packet loss, 10%, corresponds to the "normal updates"

In order to fulfill the above QoS requirements, a proposed high-level architectural design for the transport of the supermedia streams is depicted in Fig. 1.

It contains separate communication channels for each media stream. Those are as follows:

Haptic Control Channel. It transfers commands from the user to the remote haptic equipment. A specific for haptic data transport protocol should be used inside this channel. Strict QoS rules should be enforced in order to fulfill the requirements of Table 2.

Haptic Feedback Channel. It carries haptic feedback from the remote haptic interface/server back to the user. Strict QoS rules should also be enforced based on Table 2. All haptic feedback data should be synchronized with other sensor feedback data, as well as video and audio in order to maximize the QoE of the user.

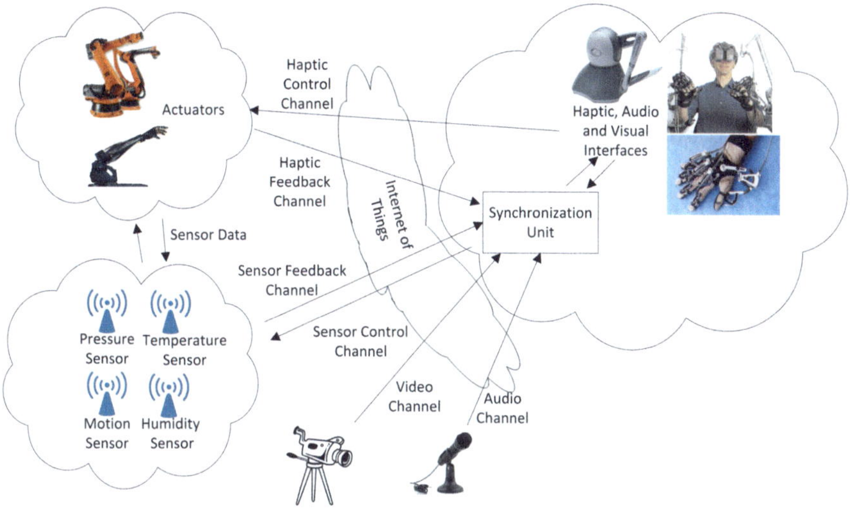

Fig. 1 High-level architectural design of the proposed supermedia system

HEVC Video Channel. It transfers a H.264/HEVC video stream. This stream has the highest throughput from all other streams. Real-time protocols as RTP and UDP are usually used for this transmission.

Audio Channel. It transfers audio data from the monitoring environment back to the synchronization unit. Again RTP and UTP protocols are usually used.

Sensor Control Channel. It transfers control commands to the remote sensors. These commands should be transferred reliably, so TCP protocol is usually used.

Sensor Feedback Channel. This channel carries sensor data to the synchronization unit. This data usually needs low QoS requirements. The update rate and the packet size of this stream are usually small.

All the above channels deliver their data to the **synchronization unit**. Synchronization techniques as the virtual-time contraction and expansion, the skipping, and the shortening and extension of output duration are enforced in order to synchronize all media streams [6].

3 Flow Control Filtering Algorithms for Supermedia Transferring Over IoT

Supermedia streams demand different QoS for each media stream. In order to fulfill these QoS, different flow control techniques should be enforced in each stream. Streams with strict QoS should have higher priority than others. Moreover, since the network conditions of the IoT are time-varying, the flow control algorithms should be network adaptive. The transmission rate, the packet size, and the

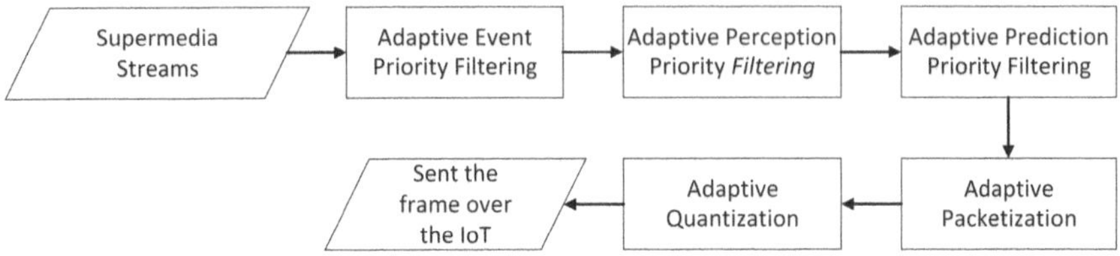

Fig. 2 System model of network adaptive flow control algorithm for supermedia streams over IoT

throughput of each stream should be network adaptive. If the network is showing signs of congestion, the transmission rate and the packet size should be reduced in order heavy congestion to be avoided. If network conditions are deteriorating, packets with lower priority should be filtered and dropped. Adaptive differential coding and quantization should modify the packet size according to the network conditions.

The proposed network adaptive flow control algorithm for supermedia streams over the IoT is depicted in Fig. 2. Each step of the algorithm is analyzed on the following subsections.

3.1 Network Adaptive Event Priority Filtering

Packets that correspond to significant events/keypoints should have higher priority than other simple update packets and should be sent more reliably [7]. When network conditions deteriorate, packets with lower priority should be buffered or dropped. Such a keypoint could be an instantaneous decoder refresh (IDR) frame of a H.264/265 video stream, or a significant haptic event, such as grasping a virtual object.

3.2 Network Adaptive Perception Priority

The perception priority is based on the Weber's law of Just Noticeable Difference (JND) [4]. Equation (1) calculates the JND. The factor I is the stimulus intense that the supermedia interface causes to the user. The dead-reckoning theory [8] supports that supermedia packets that generate small perceptible feelings ΔI to the user should be dropped. The constant κ is called the Weber fraction.

$$\Delta \mathrm{I} = I^{*}\kappa \tag{1}$$

When the supermedia packets produce to the user difference on the stimulus intense dI higher than the threshold ΔI, then the packet should obtain high priority. The higher the dI is, the higher the priority should be.

In order for the algorithm to be network adaptive, the Weber fraction κ should change according to the network conditions. The constant κ should increase when network conditions deteriorate, in order to lower the throughput of the stream. The bigger the Weber fraction is, the smaller the QoE of the user.

3.3 Network Adaptive Prediction Priority

Packets that can be predicted by previous packets should obtain a lower prediction priority [9]. The prediction unit is installed both at the sender and at the receiver. If the prediction unit at the sender calculates that the current packet could be predicted at the receiver from the last packets that were send, then the current packet is not transmitted. In order for the algorithm to be network adaptive, apart from the identical predicted packets, predicted packets that are similar to the real packets could be excluded from the transmission. Again this algorithm should be based on Weber's law of the JND to decide which packets could successfully be predicted at the receiver side. If the predicted packet does not produce greater difference on the stimulus intense dI to the users from the real packet than the threshold $\Delta I'$, then the packet is not transmitted, but it will be predicted on the receiver's side.

3.4 Network Adaptive Transmission Rate

The source should adapt its transmission rate according to the network conditions [10]. If the supermedia interface produces update packets steadily and the sender fluctuates the sending rate, a buffer is necessary at the sender side to absorb the fluctuation. The negative aspect of this technique is that if the supermedia interface produces packets at very high update rate, for haptics is usually 1 kHz, the sender should transmit its packets sometimes even faster to compensates the previous lower rates. Such a high update rate often results in network congestion and packet loss. In order to lower the update rate, an interesting proposal is to integrate a group of packets in a frame and send them as a unified packet, a technique called packetization interval. Fujimoto and Ishibashi [11] have proven that a packetization interval with eight packets per frame that is sent every 8 ms instead of 1 packet/s improves system performance in overloaded networks. This means that the sending rate could be lowered to the 1/8 of the 1 kHz.

3.5 Network Adaptive Quantization and Differential Coding

The bandwidth that a supermedia stream absorbs depends on two factors, the frame rate and the size of the frame. The frame size could be reduced if differential coding and quantization [12] are enforced on the packets that are grouped in the frame. The technique that is recommended is the differential pulse-code modulation (DPCM). The differential coding transmits not the hole packet but the difference between the reference point and the processed packet. Differential coding produces smaller packets than the original ones. Smaller packets mean fewer bits. The quantization of the differentiate values could be made with variable quantization step. The adaptive differential pulse-code modulation (ADPCM) for haptic packets was introduced in [12]. The quantization step should be changed according to the network conditions. When the network conditions deteriorate, the quantization step should increase in order for fewer bits to be required for the reconstruction of the original values.

4 Measuring Performance for High Update Rate Streams Over WiFi Repeaters

Network conditions refer to the amount of traffic that is being transferred through the Internet, the end-to-end delay, the jitter between source and destination, and the available bandwidth for data transport.

All the above metrics vary in time and space. They depend on the number of the online users, the amount of data that is being transferred at the specific moment of the measurement, and the available equipment of lines and routers. The amount of data transferred through the Web is constantly increasing. The number of the online users is increasing as well. The growth of data transfer is compensated by continuing infrastructure upgrades of computer networks.

There are two types of approaches for monitoring the network status. The two disciplines of network monitoring are the active and the passive measurements [13]. In the active measurement, specific generated probe packets, ICMP messages, are sent to specific destinations; measurements for delay, round-trip time, jitter, and packet loss are made. Some common diagnostic tools for active measurements are the ping, traceroute, capprobe, pathchar, netem, and dummynet [14]. On the other hand, passive approach is based on the observation of the traffic that flows on the links. Some passive monitoring tools, commonly called sniffers, are the Tcpdump, Wireshark, Ethereal, Netflow, and JFlow [15].

The authors actively measured the average, the standard deviation of the delay, and the packet loss rate in a WiFi multihop network with wireless access point repeaters. Two different topologies were used for these measurements. The first scenario was a simple WiFi network with one access point (AP). The second scenario was a WiFi network with one wired AP and one wireless AP repeater. The access points that were used were the 300 Mbps Tenda A30 Wireless N300 Range Extender. The packet size and the update rate of the streams were changing in order to examine whether the transferring of high update rate of supermedia streams through wireless multihop networks is possible. Two different update rates were used in all network topologies, a stream with 1000 packets per sec and a stream

Table 3 Network status for supermedia streams

Network	Update rate (Packets/sec)	Packet size (Bytes)	Round-trip time (ms)	Standard variation of RTT (ms)	Packet loss (%)
WiFi with no AP Rep.	1000	64	3.65	6.19	0.06
WiFi with no AP Rep.	1000	128	3.8	6.59	0
WiFi with no AP Rep.	500	128	3.21	5.15	0
WiFi with no AP Rep.	500	256	3.78	6.18	0.07
WiFi with one AP Rep.	1000	64	11.18	8.85	0.39
WiFi with one AP Rep.	1000	128	14.97	17.98	0.8
WiFi with one AP Rep.	500	128	7.37	8.20	0.4
WiFi with one AP Rep.	500	256	10.98	15.16	0.42

with 500 packets per sec. The packet size of the stream was changed from 64 bytes to 128 and 256 bytes. The results of these measurements are shown in Table 3.

It is obvious that even the round-trip delay of Table 3 is by far smaller than the QoS requirements regarding the delay of Table 2. On the other hand, the jitter of Table 3 is bigger than the jitter QoS requirement of Table 2. This exceedance can be limited if a network adaptive buffer is placed at the receiver side to absorb the delay fluctuation and diminish the jitter.

Furthermore, it is understood that when the update rate increases and the packet size remains steady, the network delay, the jitter, and the packet loss are increased as well. This means that the network conditions deteriorate. Similarly, when the packet size is increased and the update rate remains the same, the network conditions are deteriorating as well.

An interesting case is when the update rate is minimized to half, from 1000 to 500 packets/s, and the packet size doubles its size, from 128 bytes to 256 bytes. This means that the throughput of the application remains the same. The network conditions in this case are improved. This improvement is more obvious when the wireless network is consisted from wireless AP repeaters. The proposed flow control algorithms take advantage of this observation and try to lower the sending rate of the stream by grouping the packets into bigger frames, especially when network conditions are not adequate. Moreover, compression techniques as differential coding and quantization are enforced to these frames in order to minimize the transferred frames. From Table 3, it is understood that lower sending rates and smaller transferred packets minimize delay, jitter, and packet loss.

5 Conclusions

Wireless networks are an integral part of the IoT. The Internet service providers should integrate the network conditions that are required for the transfer of supermedia streams in their QoS in order for the IoT to flourish. The 5G mobile network is promising as its main goal is to offer bandwidth more than 1 gigabits per second, latency lower than one millisecond, and energy efficiency higher than the 4G.

The experiment tests that have been carried out in this paper proven that the transfer of real-time high update rate supermedia streams over IoT wireless multihop networks is challenging. In wired networks, the network conditions are sufficient for high update rate transferring. In wireless multihop networks, high update rates cause increased end-to-end delay, jitter, and packet loss. This increase is observed especially when the update rate is high, or when the transferred packet size is getting bigger or when the hops in the wireless network are multiplied. In order to minimize these effects, a flow control algorithm for high update rate supermedia data transferring over wireless multihop networks is proposed in this paper. Prediction, compression, packet priority, and filtering techniques are enforced in order to minimize the update rate and the packet size of the supermedia streams. Applying the proposed flow control delay, jitter and packet loss are reduced.

References

1. INFSO D.4 Networked Enterprise & RFID INFSO G.2 Micro & Nanosystems, in *Co-operation with the Working Group RFID of the ETP EPOSS, "Internet of Things in 2020, Roadmap for the Future"*, Version 1.1, 27 May 2008
2. Z. Sheng, S. Yang, Y. Yu, A. Vasilakos, J. McCann, K. Leung, A survey on the IETF protocol suite for the internet of things: standards, challenges, and opportunities. IEEE Wirel. Commun. **20**(6), 91–98 (2013)
3. K. Iwata, Y. Ishibashi, N. Fukushima, S. Sugawara, Qoe assessment in haptic media, sound, and video transmission: effect of playout buffering control. Comput. Entertain. (CIE) **8**(2), 12 (2010)
4. S. Allin, Y. Matsuoka, R. Klatzky, Measuring just noticeable differences for haptic force feedback: implications for rehabilitation, in *Proceedings of the 10th Symposium on Haptic Interfaces for Virtual Environment and Teleoperator Systems*. IEEE (2002), pp. 299–302
5. S. Dodeller, N.D. Georganas, Transport layer protocols for telehaptics update message, in *Proceedings of the 22nd Biennial Symposium on Communications, Queen's Univeristy, Canada*, May 31–June 3 (2004)
6. Q. Zeng, Y. Ishibashi, N. Fukushima, S. Sugawara, K. Psannis, Influences of inter-stream synchronization errors among haptic media, sound, and video on quality of experience in networked ensemble, in *Proceedings of the IEEE 2nd Global Conference on Consumer Electronics (GCCE)*, Oct 2013, pp. 466–470
7. S. Lee, J. Kim, Priority-based haptic event filtering for transmission and error control in networked virtual environments. Multimed. Syst. **15**(6), 355–367 (2009)

8. Y. Ishibashi, Y. Hashimoto, T. Ikedo, S. Sugawara, Adaptive delta-causality control with adaptive dead-reckoning in networked games, in *Proceedings of the 6th ACM SIGCOMM Workshop on Network and System Support for Games*. ACM (2007), pp. 75–80
9. C.W. Borst, Predictive coding for efficient host-device communication in a pneumatic force-feedback display, in *Proceedings of the Eurohaptics Conference, Symposium on Haptic Interfaces for Virtual Environment and Teleoperator Systems*. IEEE (2005), pp. 596–599
10. R. Wirz, R. Marn, M. Ferre, J. Barrio, J.M. Claver, J. Ortego, Bidirectional transport protocol for teleoperated robots. ÍEEE Trans. Ind. Electron. **56**(9), 3772–3781 (2009)
11. M. Fujimoto, Y. Ishibashi, Packetization interval of haptic media in networked virtual environments, in *Proceedings of the 4th ACM SIGCOMM Workshop on Network and System Support for Games*. ACM (2005), pp. 1–6
12. C. Shahabi, A. Ortega, M.R. Kolahdouzan, A comparison of different haptic compression techniques, in *Proceedings of the IEEE International Conference on Multimedia and Expo, ICME'02*, vol. 1. IEEE (2002), pp. 657–660
13. A. Callado, C. Kamienski, G. Szabo, B. Gero, J. Kelner, S. Fernandes, D. Sadok, A survey on internet traffic identification, in *Communications Surveys Tutorials*, vol. 11, no. 3. IEEE (2009), pp. 37–52
14. A. Finamore, M. Mellia, M. Meo, M.M. Munafo, D. Rossi, Experiences of internet traffic monitoring with Tstat. IEEE Netw. **25**(3), 8–14 (2011)
15. A. Callado, C. Kamienski, G. Szabo, B. Gero, J. Kelner, S. Fernandes, D. Sadok, A survey on internet traffic identification. IEEE Commun. Surv. Tutor. **11**(3), 37–52, 3rd Quarter (2009)

Smart Connectivity for Internet of Things (IoT) Applications

Albena Mihovska and Mahasweta Sarkar

Abstract The IoT scenario is characterized with ultra-dense interworking of billions and billions of devices through a myriad of technologies for the delivery of smart personalized services and applications. The main challenges and distinctive features of this scenario are the large amount of information gathered from the ambient environment and the human body that must be processed mostly in real- or near-real time for the unobtrusive delivery of personalized and often of critical to the user's well-being services. Smart connectivity in an AAL context relates to the availability of a reliable data channel between devices and between the human and devices and enabling an interface to the cloud/network where information gets personalized. This chapter examines the main research challenges related to the enabling of smart connectivity in an AAL context and proposes a novel approach to scalable and autonomous interactions for an enhanced personalized experience.

Keywords Smart connectivity • Internet of things • Real-time communication • Smart environments • User-centric IoT applications

The use of information and communication technologies (ICT) has brought numerous benefits to society, nominating it as a major driver for innovative smart and assisted living (AAL) applications [1–5]. This trend is backed by the rapid development of the concept of the Internet of Things (IoT) that is a key enabler of a smart ambient environment.

The IoT scenario is characterized with ultra-dense interworking of billions and billions of devices through a myriad of technologies for the delivery of smart personalized services and applications. The main challenges and distinctive features

A. Mihovska (✉)
Aarhus BSS, Aarhus University, Birk Centerpark 15, Herning, Denmark
e-mail: amihovska@btech.au.dk

M. Sarkar
Department of Electrical and Computer Engineering, San Diego State University, San Diego, CA, USA
e-mail: msarkar2@mail.sdsu.edu

R.R. Yager and J. Pascual Espada (eds.), *New Advances in the Internet of Things*, Studies in Computational Intelligence 715, DOI 10.1007/978-3-319-58190-3_7

of this scenario are the large amount of information gathered from the ambient environment and the human body that must be processed mostly in real- or near-real time for the unobtrusive delivery of personalized and often of critical to the user's well-being services.

Smart connectivity in an AAL context relates to the availability of a reliable data channel between devices and between the human and devices and enabling an interface to the cloud/network where information gets personalized.

This chapter examines the main research challenges related to the enabling of smart connectivity in an AAL context and proposes a novel approach to scalable and autonomous interactions for an enhanced personalized experience.

This chapter is organized as follows. Section 1 describes the trends in smart connectivity in relation to the emergence of the concept of user centricity and related scenarios. Section 2 describes the enabling technologies for smart connectivity and identifies the main research challenges to overcome for enabling smart connectivity deployment. Section 3 defines a scenario for ultra-dense smart connectivity and identifies the critical requirements related to its real-life deployment. Section 4 concludes this chapter.

1 Road map of Smart Connectivity

Pushed by the user demand of data availability while on the move, the worldwide research has been undergoing rapid developments observed since the millennium in the areas of wireless short-range, cellular, and satellite technologies, on the one hand, and computing and artificial intelligence, on the other. This, in turn, has opened new business horizons, which evolved the concept of IoT and brought forward the user as a center point for content generation and business revenues. Both, the desires to stay informed globally, as well as to enhance the quality of life, have made possible that social interactions have gained a new dimension, not only related to the mere communication within the immediate user-centric circle but also to the possibility of benefiting, both, user and society in the broader aspect of improved socioeconomics. There is a strong interrelationship between the growth in devices, the user demand for data, and the research development of enabling wireless technologies. This is shown in Fig. 1.

In this context, end user services and applications get a determining role for the research direction in technologies enabling smartness, which are evolving to allow for new ways of utilizing existing and building new infrastructure, and for enhancing the socioeconomic life of a city, region, country, or worldwide [6]. This chapter deals with contextualizing the role of smart connectivity for the delivery of user-centric ICT applications, such as eHealth and AAL.

Some of the most prominent technologies emerging today as key enablers of smart user-centric ICT applications are sensor technology, short-range wireless and indoor localization, cloud and fog computing, networking, information and machine learning technology, decision support systems, modeling of behavioral and

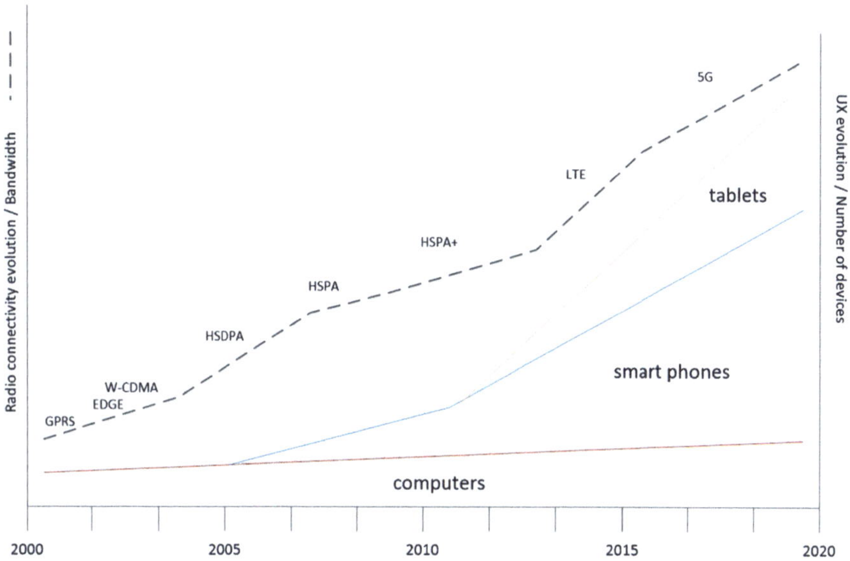

Fig. 1 Mutual influence of growth of number of devices, user demand, and wireless technologies development

cognitive processes, as well as system and process modeling. This concept has also become known as human-centric IoT [7] and brings in many implications to research to realize a vision of complex and ultra-dense connectivity. To realize a human-centric IoT scenario, smart connectivity is required at several levels as shown in Fig. 2.

1.1 Smart Connectivity for User-Centric Scenarios

A number of user-centric scenarios enabled by smart connectivity have emerged in the last decade. The most prominent ones that have been receiving worldwide research and business interest are eHealth, smart city, smart grid, smart vehicles, and a few others.

Mihovska et al. [7] introduced the concept of the smart body area network (S-BAN) [8] as the smallest component within a user-centric scenario. The S-BAN enables an interface between the human with a device or a network of sensors. The S-BAN is a key element of the human-centric connectivity as shown in Fig. 2 and is motivated by applications related to eHealth or simply for daily activity and safety monitoring. Data communication is enabled by means of radio propagation channels and networks that are located on the human body [9–11].

The S-BAN can connect to other S-BANs, devices, and networks of sensors in the environment by means of smart wireless connectivity. This scenario can be

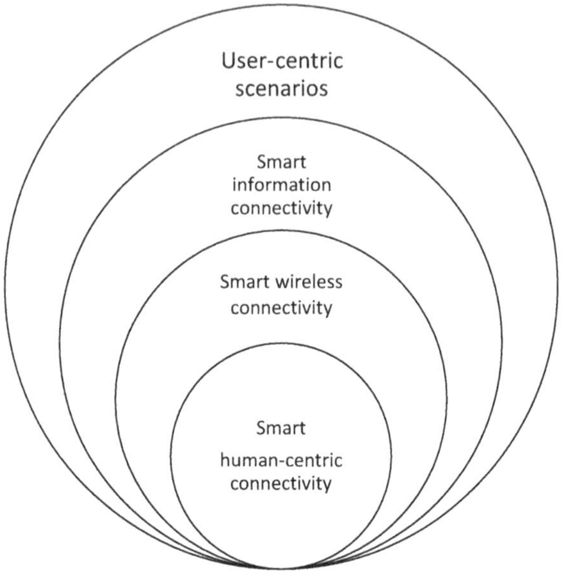

Fig. 2 Smart connectivity for human-centric IoT

explained better with the example of in-home monitoring. In-home monitoring refers to the process of acquiring sensor and audiovisual data from the home user environment and applying techniques such as machine learning, fuzzy logic, and other artificial intelligence algorithms to make sense out of it. A key prerequisite here is that this connectivity is enabled anytime and anywhere in order to realize more complex user-centric scenarios, for example, deriving information from the ambient environment or for personal safety (e.g., fall detection) [12]. In this situation, data communication will be enabled by means of short-range wireless technologies, and this information would need to be processed in real time, which requires to have in place technologies such as fog computing [12].

The smart information connectivity (see Fig. 2) enables the delivery of personalized services within a user-centric scenario by connecting to the cloud and relying on technologies such as lifestyle reasoning, semantics, and intelligent decision support systems (IDDS), to make decisions about short-, medium-, and long-term human-centric situations [13]. Analytics and big data are essential elements here.

The most essential feature of smart connectivity for user-centric scenarios is the unobtrusiveness and seamlessness of the technology. These features in turn require that security, trust, and privacy procedures are properly put in place to guarantee user-friendliness, the protection of personal user data, and the safety of the user.

It may be expected that the degree of achievable intelligence within a user-centric scenario will allow for communication capabilities for any type of 'thing' and not only the ones equipped with a traditional radio electronics [14]. This concept has seen an increased research interest into non-radio connectivity

Fig. 3 VLC-enabled smart connectivity

technologies, such as visible light communication (VLC) [15]. A VLC-enabled smart connectivity scenario is shown in Fig. 3.

In the scenario of Fig. 3, smart connectivity at the level of the first two circles of Fig. 2 will be enabled by the seamless handover between VLC and short-range radio technologies. VLC is enabled by an ambient lighting system (ALS) and comprises a series of light-emitting devices (e.g., the recently emerged light emission diodes—LEDs) that can be controlled by a single or multiple switches for illuminating the target area with visible light [15]. Because this technology still faces the open challenges of line-of-sight path and the need to switch on the light, in certain contexts, where the above conditions are not present, the connectivity can be established by means of Wi-Fi, Bluetooth Smart (i.e., Bluetooth Low Energy), and similar. The advantages of introducing VLC are a relieved radio spectrum, which in the context of a growing number of IoT connections and devices, is a critical issue to consider. A detailed proposal of how to integrate VLC with the existing Wi-Fi and cellular infrastructure is provided in [15].

A large portion of the research and standardization effort is currently in the direction of enabling direct device-to-device communication, in particular in an AAL scenario. One example of such a standard is Miracast, which allows a mobile device, for instance, to discover and connect to another device, such as a television, in order to mirror the contents of its screen to the television's display [16].

Smart connectivity for the realization of user-centric scenarios is strongly dependent on the ability to enable interoperability at the various levels of the connectivity process. The next Sect. 2 analyzes the key enabling technologies to emphasize on the open research challenges toward true smart connectivity.

2 Enabling Technologies and Open Challenges

The IoT concept has been designed around the capability to perform several main distinctive actions, namely the following:

- Sensing;
- Data collection;
- Data transmission/exchange;
- Data processing;
- Data storage;
- Data personalization.

The above require a number of functionalities that, in general, will be implemented as physical and logical entities, joint together in a platform (i.e., service-oriented platform-SoA) [17]. An example of functionalities within an SoA platform capable of supporting personalized eHealth applications as an IoT user-centric scenario is shown in Fig. 4. Effective solutions must satisfy a multitude of constraints arising from clinical/medical needs, social interactions, cognitive

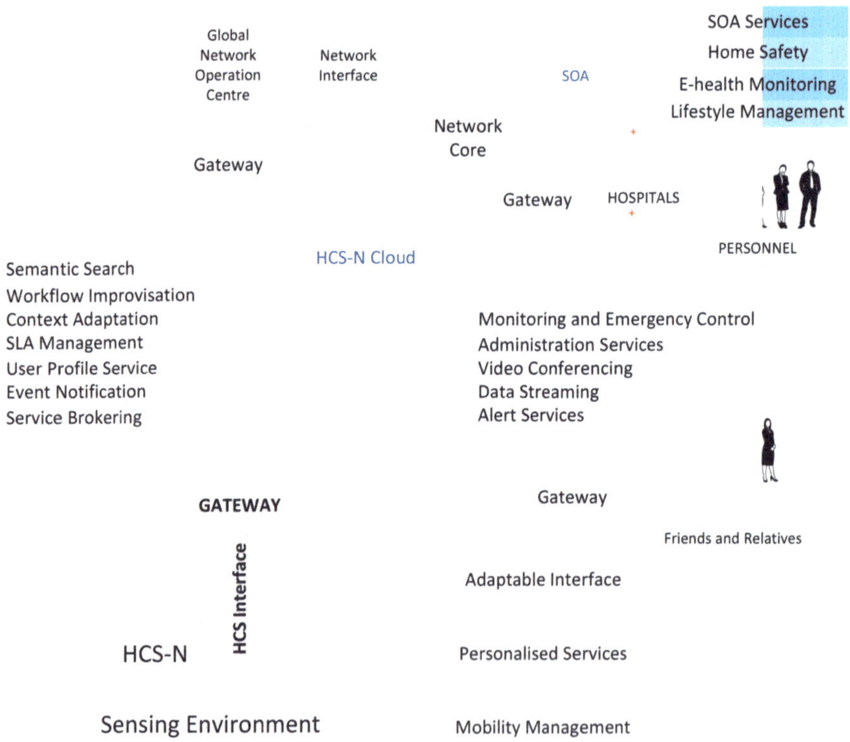

Fig. 4 eHealth user-centric scenario [18]

limitations, barriers to behavioral change, heterogeneity of data, semantic mismatch, and limitations of current IoT systems.

2.1 Platforms

The scenario in Fig. 4 is centered around two categories of end users, namely the primary user in the home, who needs medical care, and the secondary users, who take care of the primary user. The services and applications are personalized to cater for the needs of the primary user. The common platform, enabling the connectivity between the two types of end users and the end users and applications is realized as a service-oriented architecture (SOA) approach in order to create an architecture based upon the use of services. The main research environments associated with this scenario are the sensing environment, which enables the collection of data from the user's body or/and ambience, and forwarding these data via a gateway to the cloud environment for further processing to enable the provision of personalized services. The goal is to create modular, adaptive, unobtrusive, reliable, motivating, and easy-to-use system. To this end, a number of functionalities are needed to satisfy the following mandatory system requirements [19]:

- Interoperability—ability to interconnect with other systems (e.g., epSOS);
- Security—ability to secure the users' data from obtrusive and accidental eavesdropping;
- Privacy—ability to keep personal information from being disclosed and shared with unauthorized parties;
- Context information—ability to provide context information that is useful for services to adapt themselves according to the needs, preferences, and situation of the user;
- Service orientation—ability of a system to ensure reusability and composability of services and service components;
- Semantic interoperability—ability to enable semantic interoperability between applications and services for ensuring the highest degree of decoupling (enables an open system and facilitates reuse of existing services and applications);
- Maintainability and configurability—ability to easily maintain and configure the system after deployment;
- User data separation—ability to create pseudoidentifiers for privacy protection supported by the relevant user and network devices;
- Distributed decision making—ability to make decisions in user's home system in addition to the user's cloud environment;
- Anonymity—ability to switch off the sensors and devices and manage the deletion of raw data from these sensors.

In addition, some specific system requirements would be needed to satisfy the various capabilities of the implemented technical components. Examples of some mandatory specific system requirements are given below:

- Ambient parameters monitoring—ability to monitor critical ambient parameters such as temperature, air quality, and smoke;
- Monitoring users' vital parameters—ability to monitor (using wearable sensors) body temperature, and blood pressure;
- Monitoring users' associated vital parameters—ability to monitor vital parameters that affect the users' status such as sleep length, sleep quality, and amount of conducted physical activity;
- Reasoning—ability to reason upon stored user behaviors and choose appropriate actions (e.g., reminders for taking prescribed medicine);
- Adaptive A/V formats—ability to support audio and video bit rate adaptation to accommodate various possible display devices;
- Remote accessibility—ability to provide remote access to the platform;
- Identification, authentication, and authorization—ability of system components to identify, authenticate, and authorize an entity (human users and other system components) that wants to use them before allowing them access to resources;
- Confidentiality—ability to maintain confidentiality (the way in which the information disclosed or managed by the system is treated) of identifiable data, including controls on storage, handling, and sharing of data;
- Integrity—ability to detect data modifications and prevent unauthorized modifications, especially related to service user data, sensor data, and commands sent to actuators;
- Non-repudiation—ability to trace back every action on sensitive assets to the person or system component that performed it;
- Auditing—ability of a system to log all actions on sensitive assets, including failed access attempts;
- Consent specification—ability to provide a usable interface to capture the consent of the end user about sharing data with services;
- Communication—ability to enable intercomponent message-based (or event-based) and call-based communications between distributed components.

2.2 *Protocols and Interfaces*

An important part of a successful IoT platform is to carefully define the interface requirements, which in turn are very important for defining the communication protocols, protocol formats, and messages. A key current challenge for realizing successfully an IoT scenario is the device and network protocol connectivity and interoperability. Protocols are crucial for enabling the added value of big data for the delivery of personalized services. It is important that support for legacy protocols and new protocols, and bringing those together under an interoperability umbrella is put into place, and a big effort is put into this by international standardization bodies, such as the International telecommunication Union (ITU) [20], the European Telecommunication Standardization Institute (ETSI) [21], and the Telecommunication Industry Association (TIA) [22]. Protocol interoperability and device management should ensure that future devices will easily fit into any platform implementation scheme, which would enable backward compatibility of already developed platforms.

For the scenario in Fig. 4, at least the following interfaces must be put into place to enable the proper platform operation [19]:

- Interface between the sensing environment and the cloud: to facilitate the communication and information distribution of the congregated information from the home environment sensors toward the cloud middleware and to disseminate the control and configuration data from the cloud middleware toward the sensing environment.
- Interface to leverage the information exchange between the cloud middleware components (i.e., data manager and cloud database.) and the platform services (i.e., profiling servers, IDSS, and notification manager). This is an internal cloud interface between the cloud environment components that supports the distribution and exchange of high-level data formats.
- Interface to facilitate low-level data (i.e., raw sensor data) and high-level data (i.e., processed data) exchange between the cloud middleware entities (e.g., data manager and cloud database) and the service bricks.
- Interface between the cloud environment and external sources (e.g., epSOS in Fig. 4). The interface provides bidirectional information data exchange between the cloud components (such as the data manager and the data fusioner) and the external systems and supports both datagram-based and streaming communication.
- Interface to facilitate high-level data exchange between the platform services (e.g., IDSS and notification manager) and the service bricks in the form of processed data, as well as high-level data exchange between the platform services and applications in the form of semantic metadata (i.e., IDSS and lifestyle reasoners decisions, alarms, and reminders.)
- Interface to leverage the communication between the service bricks and the applications for the distribution of specific aggregated metadata information from the service bricks toward the user-centric applications.

2.3 *Low-Power IoT Design*

One key prerequisite for successful IoT platform operation is related to power consumption and energy efficiency of the platform design, which would directly impact the overall application performance, such as latency. Low-power IoT design would require that on the one hand, novel battery solutions are put into place

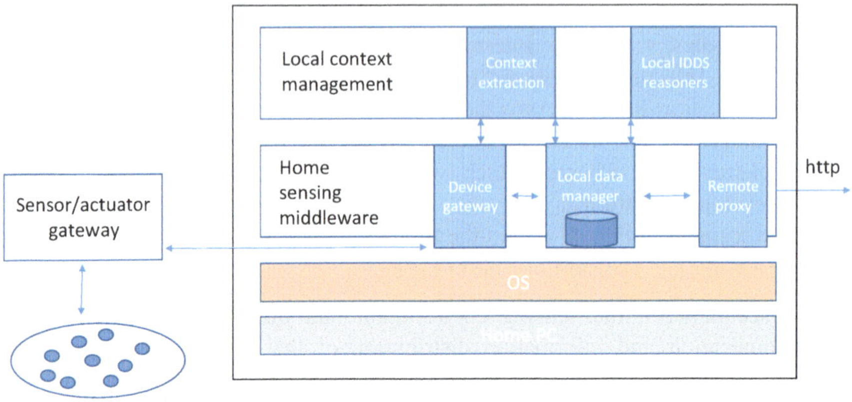

Fig. 5 Logical deployment of a home sensing environment [19]

(e.g., solar power), and on the other that an IoT platform solution is tested end-to-end under real-life network conditions and based on a test solution that allows very precise power and current measurements.

An example of a logical deployment realization of the sensing environment in Fig. 4 is shown in Fig. 5.

A set of physical sensor devices are distributed in the home, together with a dedicated device which acts as a gateway that collects signals from the sensors. A set of software components, whose purpose is to collect, process, store, and transmit sensor data, are distributed together on a common runtime environment. The successful component communication will be very dependent on the energy efficiency of the signaling and the chosen hardware components. It must be noted that a large portion of the data collected from the sensing environment would require real- or nearly real-time processing, which can be power-consuming, and, therefore, the processing device (e.g., the gateway in Fig. 4) should be adequately chosen. Another issue to consider is that for an IoT system to be successful, it should have an open nature, and not rely on proprietary technologies. Therefore, an end-to-end energy-efficient and low-power design must be in place.

2.4 *Interference-Free Smart Connectivity*

IoT smart connectivity enables the collection and transmission of data to the cloud without human intervention. IoT smart connectivity is driven by the fact that most of the devices would be operating in an unlicensed band and will have the so-called 'cognitive' capability to utilize available spectrum holes. Current IoT connectivity standards include communication protocols such as Bluetooth, Wi-Fi, Zigbee, and USB and allow for deploying an IoT scenario with off-the-shelf and very low-cost components, which also get increasingly miniaturized. However, the connectivity and sensing between the IoT device and the human, which is also the core of the human-centric concept, remains a challenge. Open research issues for enabling reliable IoT connectivity are toward RF interference mitigating technologies, energy-efficient short-range technologies, new channel models to account for the specifics of the human body medium, and their effect on the radio transmissions. Current research in the context of an S-BAN connectivity based on implantable devices has been investigating the ultra-wideband (UWB) channel and associated interference issues related to the particularities of the human tissue medium [23, 24].

In the context of ultra-dense connectivity that is associated with IoT scenarios, wireless interference can severely degrade the performance and jeopardize proper service delivery, which in the context of user-centric scenarios, such as eHealth, smart grid, smart car can be of lifesaving importance. In a smart home scenario, even everyday household appliances could be source interference (e.g., microwave ovens). Earlier, in this chapter, we mentioned the research trend to combine seamlessly RF and non-RF (e.g., VLC and mm-wave) technologies in an ambient scenario in order to accommodate an ever growing number of connections and devices. Also, here, it is required that research is closely paired to standardization in order to integrate successfully any such new solutions with the existing ones and the infrastructure.

In the context of a human-centric smart connectivity, certification of the deployed devices and connections is one key prerequisite for enabling user-friendly and safe solutions.

3 Smart and Connected AAL

Figure 6 shows a scenario of ultra-dense smart connectivity with a complicated signal environment. A variety of wireless technologies must coexist interference-free to deliver digital services to the users in a variety of contexts and sub-scenarios, namely smart home, smart car, smart grid, and smart city. In addition, the majority of these services are expected to be personalized.

In the scenario of Fig. 6, the main goal and challenge is to enable fast, reliable, and secure connections. Techniques, such as coverage predictions, positioning, and localization are a strong tool to assist planning and optimization of the wireless networks [25]. Typical interference analysis will use parameters such as signal strength, received signal strength indicator (RSSI), and signal ID, interference mapping to detect any anomalies, and techniques such as self-organization have been proposed to reorganize the wireless nodes for seamless connectivity provision [26].

The scenario in Fig. 6 is characterized by a connectivity dynamics that must be enabled under various wireless channels and networks for different communication purposes and that may occur over the same period of time. Smart connectivity in the context of the scenario in Fig. 6 would mean capabilities for the automatic

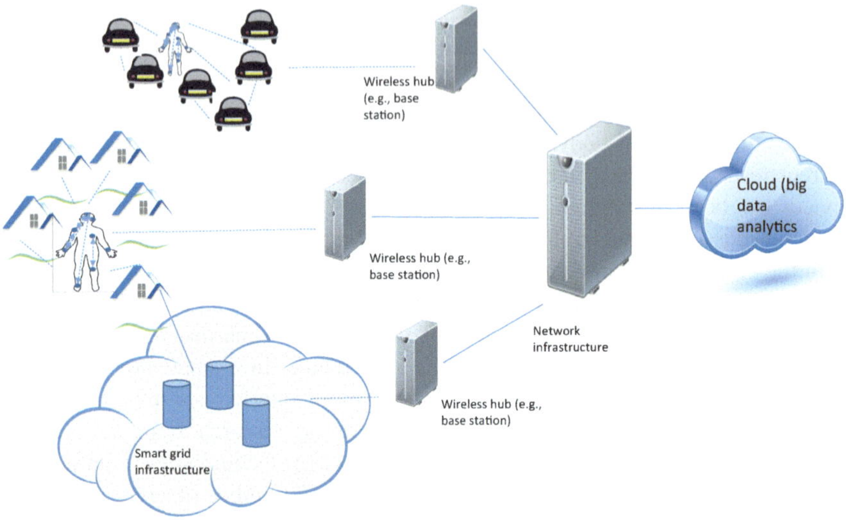

Fig. 6 User-centric smart connectivity scenario

management, identification and, use of a large number of heterogeneous physical and virtual objects (e.g., both physical and virtual representations), which are Internet-enabled and maybe on various locations. A lot of the data would be generated and collected from sensors and would be delivered it to a dedicated storage area. Fog computing allows for processing and storing real-time sensor data required to trigger alarms and similar applications requiring fast response, while the rest of the metadata streams can be processed in the cloud. The applications that may be delivered by the connectivity in Fig. 6 may involve control and monitoring functions, where human-in-the-loop actions are not required, which further tightens the security, privacy, and reliability requirements.

In addition, a large part of the established connections will be autonomous and their formation will be based on local decisions determined by the user needs. Smart connectivity must be responsive to the user-centric needs because the majority of the connections would be established for delivery of human-centric applications.

The amount of data that would be generated in the scenario of Fig. 6 would be user-centric for the most part, and very large. Big data analytics has recently began to emerge as a powerful tool to handle the various amount of information efficiently, but also, to serve as a way to optimize connectivity and reliability support. Big data analytics includes data aggregation, processing, and storage of user-centric information and has opened up even more profoundly the research challenge of how to protect these data adequately because of the societal and economic implications of any potential breaches.

4 Conclusions

This chapter explored the notion of user-centric smart connectivity in the context of IoT-based applications. Smart connectivity is the ability to provide user-centric (i.e., personalized) applications in a very complex scenario.

There are many open challenges that require new approaches to modeling of the smart connectivity network, and open and interoperable solutions for realizing a smart connectivity scenario. Security, privacy, and trust gain a new dimension of importance due to the severe implications to both the individual user as well as the society if data is breached.

Research should progress hand in hand with standardization activities to reach solid interoperable solutions that are key to smart connectivity deployment. Certification of products should be a must in order to ensure the safety of the users.

References

1. F. Sadr, Ambient intelligence: a survey. ACM Comput. Surv. **43**(4), 36:1–36:66 (2011)
2. D. Cook, J. Augusto, V. Jakkula, Ambient intelligence: technologies, applications, and opportunities. Perv. Mobile Comput. **5**(4), 277–298 (2009)
3. S. Kyriazakos, M. Mihaylov, B. Anggorojati, A. Mihovska, R. Craciunescu, O. Fratu, R. Prasad, eWall: an intelligent caring home environment offering personalized context-aware applications based on advanced sensing. Springer J. Wireless Pers. Commun. **87**(3), 1093–1111 (2016)
4. N. Zaric, A. Mihovska, M. Pejanovic-Djurisic, Ambient assisted living systems in the context of human centric sensing and IoT concept, in *2016 Proceedings of IEEE PIMRC, September 2016*, Valencia, Spain
5. A. Mihovska, S. Kyriazakos, R. Prasad, eWall for active long living: assistive ICT services for chronically Ill and elderly citizens, in *2014 IEEE International Conference on Proceedings of Systems, Man and Cybernetics (SMC)*. (IEEE Press, 2014) pp. 2204–22 09
6. National ICT Australia Report. Chapter 2: What is smart infrastructure? http://www.aph.gov.au Accessed 2017
7. A. Mihovska, R. Prasad, M. Pejanovic-Djurisic, Chapter 5: human centric IoT networks. in *Human Bond Communications*, ed by S. Dixit (Wiley, 2017)
8. ETSI Smart-BAN, Draft V0.1.0 (2015-10). Measurements and modelling of SmartBAN RF environment. Technical Report, ETSI online (2015)
9. P.V. Patel et al., Channel modelling based on statistical analysis for brain–computer-interface (BCI) applications, in *Proceedings of IEEE INFOCOM*, San Francisco, CA, April 2016
10. D.B. Smith, L.W. Hanlen, Channel modeling for wireless body area networks, in *Ultra-Low-Power Short-Range Radios, Integrated Circuits and Systems*, ed. by P.P. Mercier, A.P. Chandrakasan (Springer International Publishing, Switzerland, 2015). doi:10.1007/978-3-319-14714-7_2
11. J. Wang, Q. Wang, *Body Area Communications: Channel Modeling, Communication Systems, and EMC* (Wiley-IEEE Press, 2012). ISBN: 978-1-118-18848-4
12. R. Craciunescu, A. Mihovska, et al., Implementation of fog computing for reliable e-health applications, in *Proceedings of IEEE ASILOMAR 2015*, November 8–12, 2015, Pacific Grove, CA, USA

13. A. Mihovska, et al., eWALL innovation for smart e-Health monitoring devices, in *Wearable Technologies and Wireless Body Sensor Networks for Healthcare,* ed. by F.J. Velez, F. Derogarian (IET Publishers, 2017)
14. R. Kreifeldt, Smart connectivity. Harmann Innovation Hub, http://harmaninnovation.com/blog/smart-connectivity/. Accessed Jan 2017
15. A. Kumar, A. Mihovska, S. Kyriazakos, R. Prasad, Visible light communications (VLC) for ambient assisted living. Springer Int. J. Wireless Pers. Commun. **78**(3), 1699–1717 (2014)
16. Wi-Fi Certified Miracast, http://www.wi-fi.org/ja/discover-wi-fi/wi-fi-certified-miracast
17. Service-oriented platforms
18. EU-funded ICT project eWALL for Active Long Living (eWALL), http://ewallproject.eu
19. EU-funded ICT project eWALL for Active Long Living (eWALL), D2.7. Final user and system requirements and architecture. February 2015, http://ewallproject.eu
20. International telecommunication Union (ITU), url://itu-t.int
21. European Telecommunication Standardization Institute (ETSI), http://etsi.org
22. Telecommunication Industry Association (TIA), http://tiaonline.org
23. P.V. Patel, M. Sarkar, et al., Channel modelling based on statistical analysis for brain-computer-interface (BCI) applications, in *Proceedings of IEEE INFOCOM*, April 2016, San Francisco, CA, USA
24. ETSI Smart-BAN, Draft V0.1.0 (2015–10), Measurements and modelling of smartBAN RF environment. Technical Report (2015)
25. A. Kumar, A. Mihovska, R. Prasad, Spectrum sensing in relation to distributed antenna system for coverage predictions. Springer Int. J. Wireless Pers. Commun. **76**(3), 549–568 (2014)
26. P. Semov, V. Poulkov, A. Mihovska, R. Prasad, Self-resource allocation and scheduling challenges for heterogeneous networks deployment. Springer J. Wireless Pers. Commun. **87** (3), 759–777 (2016)

A Blockchain-Based Storage System for Data Analytics in the Internet of Things

Quanqing Xu, Khin Mi Mi Aung, Yongqing Zhu and Khai Leong Yong

Abstract Without a central authority, blockchains can easily enable the management of transactions. Smart contracts stored on blockchains are self-executing contractual states that are not controlled by anybody, so they can be trusted. In addition, due to increasing improvements in processor and memory technology, IoT (Internet of Things) devices have more powerful processing power and greater memory space, which allow them to execute user-defined programs, e.g., smart contracts. Shifting part of applications' tasks to IoT devices reduces the transferred data amount over the IoT network. The parallelism of large-scale storage systems is employed to decrease many basic data analytics tasks' execution time. Blockchain can be used as smart contracts that facilitate and enforce the negotiation of a contract in the IoT. This chapter proposes a blockchain-based storage system, named *Sapphire*, for data analytics applications in the Internet of Things. All the IoT data from the devices forms objects with IDs, attributes, policies, and methods. We present an OSD-based smart contract (*OSC*) approach employed in *Sapphire* as a transaction protocol, where IoT devices interact with such blockchains. For data analytics applications, the IoT device processors execute application-specific operations. By doing so, only the results are returned to clients instead of data files read by them. Therefore, the *Sapphire* system can greatly decrease the overhead of data analytics in the Internet of Things.

Keywords Internet of things · Blockchain · Storage system · Data analytics · Smart contract

---

Q. Xu (✉) · K.M.M. Aung · Y. Zhu · K.L. Yong
Data Storage Institute, A*STAR, Singapore, Singapore
e-mail: Xu_Quanqing@dsi.a-star.edu.sg

K.M.M. Aung
e-mail: Mi_Mi_AUNG@dsi.a-star.edu.sg

Y. Zhu
e-mail: ZHU_Yongqing@dsi.a-star.edu.sg

K.L. Yong
e-mail: YONG_Khai_Leong@dsi.a-star.edu.sg

R.R. Yager and J. Pascual Espada (eds.), *New Advances in the Internet of Things*,
Studies in Computational Intelligence 715, DOI 10.1007/978-3-319-58190-3_8

1 Introduction

The IoT (Internet of Things) is a network that is able to connect many objects to the Internet via a large number of devices, e.g., sensors, cameras, smart phones, and RFID (Radio-frequency identification) readers. All common physical objects in the IoT have an IP address or URI, and they can exchange information among them. It finally reaches a goal of intelligent management and recognition. The IoT *devices* (or *things*) are seamlessly linked into a virtual world via the IoT network, enabling anywhere and anytime connectivity. There would be 50 billion devices by 2020 as there are increasingly smart devices per person [1]. There would be 100 billion IoT connections, and the financial impact of IoT may be as much as $3.9 to $11.1 trillion on the global economy by 2025 [2]. Data in the IoT environment is from a large number of different devices. It represents billions of objects, thus it would be so extremely large that we must build a scalable distributed storage system.

IoT has received extensive attentions from both academia and industry recently, and its basic idea is to integrate the things into the Internet with provision of various services to users [3]. There are typical killer applications of IoT, such as smart home [4, 5], smart grid [6], and smart building [7]. As we see that there are increasing IoT devices, it is possible to use blockchain technologies [8] that manages numerous unspecified devices and processes, including communications and transactions among the devices. Without a central authority, blockchain technologies guarantee the security and credibility of data. Transactions among IoT devices are recorded on blockchains as smart contracts [9] and executed automatically to improve transaction efficiency greatly. For example, transactions and settlements are completed automatically irrespective of past relationships with other relevant IoT devices. Without central third parties, mutually distrustful IoT devices are allowed to transact safely in emerging smart contract systems. The decentralized blockchain [10] makes sure that IoT devices obtain commensurate compensation regardless of contractual breaches or aborts. For example, Ethereum Virtual Machine allows executing code in the form of so-called smart contracts on Ethereum [11], which is a Turing-complete decentralized smart contract system. Many companies or organizations have been building smart contract applications over Ethereum.

The rapid growth of data-driven applications shifts the nature of distributed storage systems. In object-based storage (or object storage) [12] systems, each object has a unique OID allowing a server/client to obtain it without its physical location, and all objects reside in a flat address space. An object-based storage device (OSD) manages lower-level space functions after allocating space for objects. Upper applications and users connect with many objects through APIs. Increasingly, we design a distributed storage system mainly for capacity instead of performance. Distribution and tiering are vital, and analytics applications are both essential and routine across any heat of data and any dimension. The applications primarily depend on semi-structured or unstructured data that is inexpensive and easy to create. As a consequence, the value of data management and analytics fuel the growth for data preservation in storage systems. In order to achieve the explosive storage growth, we need to remove layers of inefficiency from traditional storage system architectures and present a new method optimized for scale-out application requirements.

This chapter presents a blockchain-based distributed storage system, called *Sapphire*, as an evolution of *Gem* [13, 14], for large-scale data analytics applications in the IoT. This system is able to support diverse data-intensive applications. In this chapter, we depict the blockchain-based large-scale storage system, which is followed to put forward data analytics based on the storage system. We present an OSD-based smart contract (*OSC*) as a transaction protocol, in which IoT devices interact with such blockchains in *Sapphire*. We develop blockchain-based storage and processing techniques, in which object storage devices employ embedded processors in the devices to process apart from storing data. Direct data process in the drives can lead to a dramatic performance growth, by avoiding redundant data transfers across storage buses and networks.

The rest of this chapter is organized as follows. We introduce background and motivation in Sect. 2. We describe the system architecture of *Sapphire* in Sect. 3. We propose a location- and type-sensitive hashing mechanism, and a dynamic load balancing method in Sect. 4. In Sect. 5, we present an OSD-based smart contract (*OSC*) mechanism. In Sect. 6 IoT data analytics is presented. We summarize this chapter in Sect. 7.

2 Background and Motivation

In some applications, e.g., IoT storage, social network services, and cloud storage, object storage performs better than SAN (Storage Area Network) and NAS (Network-Attached Storage) [15] for large-scale semi-structured or unstructured data sets.

2.1 Object Storage

Object storage can easily support the explosive growth of data since we can scale-out OSDs geographically. Distributing data replicas can enhance data protection across multiple storage nodes. Object storage is an attractive solution to efficiently manage large-scale semi-structured or unstructured data sets from the Internet of Things. An object, as shown in Fig. 1a, consists of its ID, data, and attributes, which include metadata, policies (e.g., replication), methods (e.g., encryption/decryption), and user-/application-defined functions. As shown in Fig. 1b, each object has unique ID and pathname in object mapping. Object storage is able to be utilized for archiving the IoT data, e.g., sensor, camera, and smart phone data, with high compliance. Object storage systems offer the benefit of releasing storage space by enabling users to correctly differentiate data.

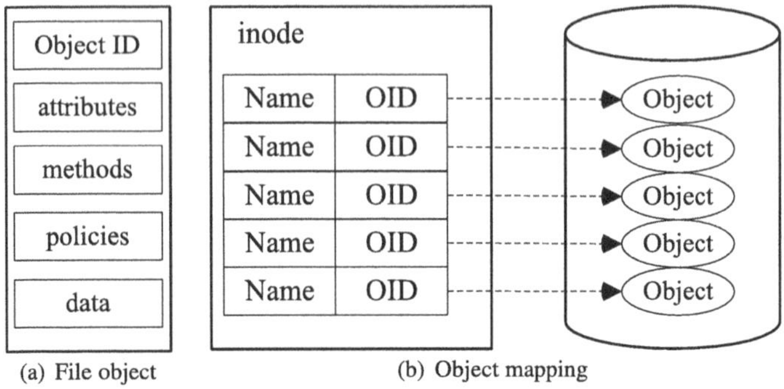

Fig. 1 Object-based storage

2.2 Requirements of IoT

The Internet of Things enables the most efficient and effective stack including systems, interfaces, protocols, and devices, to do optimizations for distributed applications. In addition, it enables object-oriented distributed applications to directly utilize storage and fuels scale-out distributed systems. In such a way, it also enables significant gains in performance and TCO (Total Cost of Ownership). The IoT data comes from a large number of different devices generating billions of data objects, and is sampled by various of perception devices, e.g., cameras, smart phones, sensors, and RFID (Radio-frequency identification) readers. However, the IoT data from different devices has distinct structures and semantics. The IoT network consists of a large number of perception devices that automatically and continuously collect information, resulting in the explosive growth of data scale. In addition, the IoT applications usually integrate a large number of sensors to simultaneously monitor many indicators, such as humidity, light, pressure, and temperature, so the sampled data is usually multidimensional. Different from traditional Internet data, the IoT data has two attributes: time and space inherently to depict dynamic state changes of object locations. Most of IoT applications are isolated, but the IoT network has to finally realize data sharing to facilitate collaborations among different IoT applications.

2.3 Smart Contract

Blockchain technology has been widely utilized by companies or organizations as a means to reorganize their centralized networks due to its decentralized nature. A blockchain, as an append-only distributed database, stores transactions that are a time-ordered set of records. Transactions are grouped into blocks and form a cryptographic hash chain in a decentralized network. Programmable electronic "smart contracts," as a conceptual idea, were proposed around twenty years ago [16]. Smart contracts are essentially automated computer programs built on a blockchain protocol, and they are made possible by general purpose calculation based on blockchains. Therefore, smart contracts can include contractual arrangement, actual execution of the contracts, and governance of the preconditions required for contractual obligations. Ethereum [11] is the first blockchain that introduces a Turing-complete scripting language with support for smart contracts. It is the most active and representative blockchain of smart contracts. Without an external trusted authority, smart contracts are software programs that are correctly executed by mutually distrusting nodes, and they are used to handle and transfer assets of considerable value in a decentralized network. Apart from their correct execution, their implementation is relatively secure against attacks that aim at tampering or stealing the assets. In order to detect some vulnerability patterns, Oyente [17] extracts and executes the control flow graph from the Ethereum Virtual Machine bytecode of a contract.

2.4 IoT Data Analytics

In [18], Wang et al. proposed DRAW including three main components: (1) data access history graph utilized to scrutinize data access patterns, (2) a data grouping matrix to organize related data, and (3) an optimal data placement method to generate final data layout. In [19], Riedel et al. proposed an active storage system, where a set of data analytics benefit from active disk drives, e.g., filtering and batching. In [20], Acharya et al. presented the idea of active storage and evaluated it in active disks. Active storage exploits the extra processing power in the drives, and it explores a stream-based programming model that allows application code to execute on the drives. In [21], Keeton et al. presented intelligent disks (IDISKs) that targeted at decision support database servers. In [22], Huston et al. proposed a concept of early discard for interactive search to filter a large amount of unindexed data for search by using a *searchlet*. Active storage concept emerged in the context of parallel file systems [23], harnessing the computing power of storage nodes, or hosts dedicated to data transfer and disk management. Therefore, active storage is excellent for IoT data analytics.

3 *Sapphire* System

In this section, we present the system architecture of large-scale blockchain-based storage system, named *Sapphire*, for data analytics in the Internet of Things.

3.1 Region Partitioning

Based on system requirements, we divide a region R covered by an IoT network into many subregions in a recursive manner [24]. Firstly, we divide the region R into two

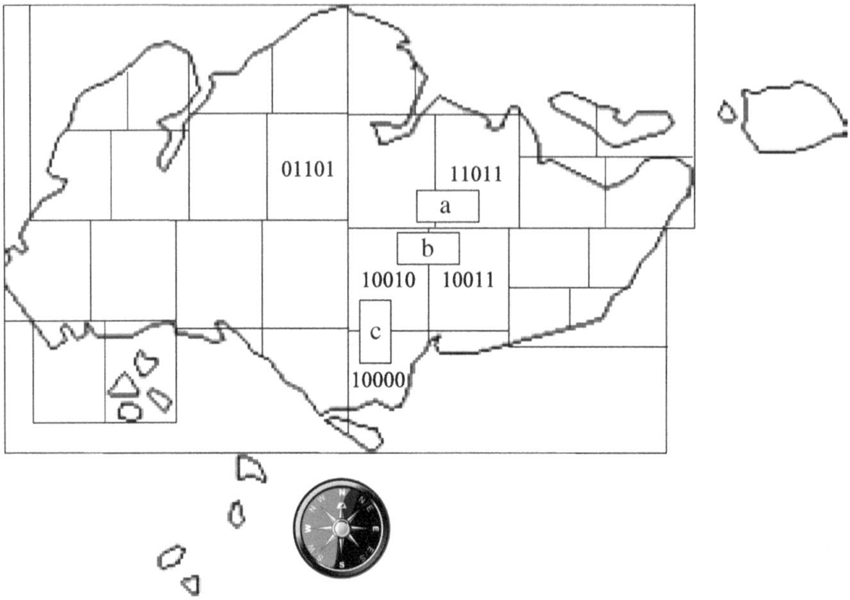

Fig. 2 Region partitioning (e.g., Singapore). Case **a** most of a building is in a region R (e.g., 11011), and this building belongs to R, case **b** half a building are in two regions R_e and R_w, and it belongs to R_w (e.g., 10010), case **c**half a building are in two regions R_s and R_n, and it belongs to R_n (e.g., 10000)

half subregions (R_e and R_w) based on longitude, where R_e and R_w are, respectively, represented by 1 and 0; and then for R_e and R_w, they are also divided into two half subregions (R_n and R_s) based on latitude, where R_n and R_s are, respectively, represented by 1 and 0 as well. Figure 2 illustrates that the above procedure is recursively processed until the differences in longitude and latitude of a subregion are both less than given thresholds *LO* and *LA*. Consequently, the entire region R is divided into multiple geographical subregions that form the network topology. Each subregion is represented by a unique ID, and each device utilizes this embedded service to keep the location information of all the subregions in the entire IoT network.

As shown in Fig. 3, we present object hierarchical namespace since the mapping from devices to subregions is known to proxy servers and all IoT devices. Folders in the object hierarchical namespace form a tree structure consisting of several levels. There are three space attributes (region, building, and device) in the first three levels, while there are two time attributes (year and month) in the last two levels. For example, all video files of a CCTV (Closed Circuit Television) camera are generated in June 2016 as shown in Fig. 3. This architecture can be used for a distributed file system (DFS), e.g., Hadoop DFS (HDFS) [25].

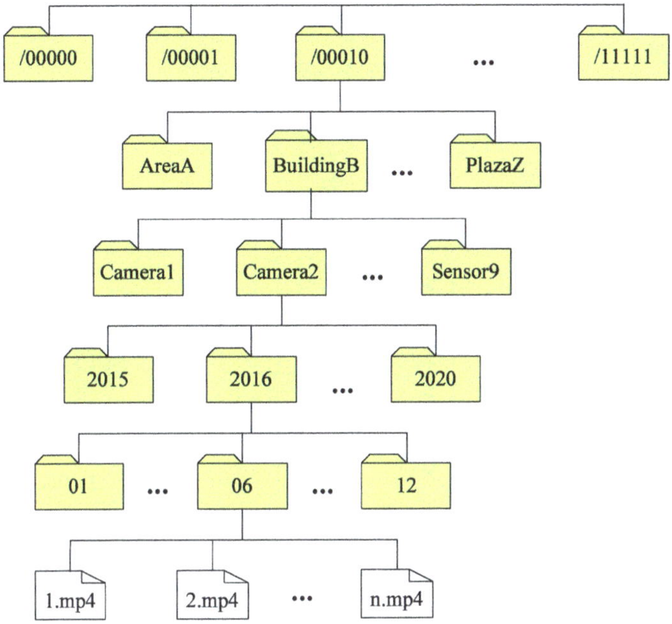

Fig. 3 Object hierarchical namespace

3.2 *IoT Device Classification*

In the *Sapphire* system, IoT devices are classified into three types: *super nodes*, *regular nodes*, and *light nodes*, according to their computational power and memory space, similar to P2P systems [26]. Super nodes have powerful computational power and large memory space, so they can manage and store complete replicas of the blockchain. The super nodes can host marketplaces and they are servers owned and deployed by companies or organizations, providing blockchain-based data analytical services and doing complex queries, and they represent the core of *Sapphire*. Indeed, they are able to perform the role of smart contracts among IoT devices across the IoT network, as long as they can balance demand and supply of services. Regular nodes are normal IoT devices equipped with regular processing power and storage space, so to meet blockchains' requirements and support light nodes, according to their capabilities. There are an increasing number of smart objects that can be included in the category of regular nodes, since the cost of chip declines. Light nodes have low resources, and they can perform messaging, transferring, and routing, but they cannot manage the blockchain, so they obtain the blockchain-based smart contracts from other trusted nodes, e.g., super nodes and regular nodes.

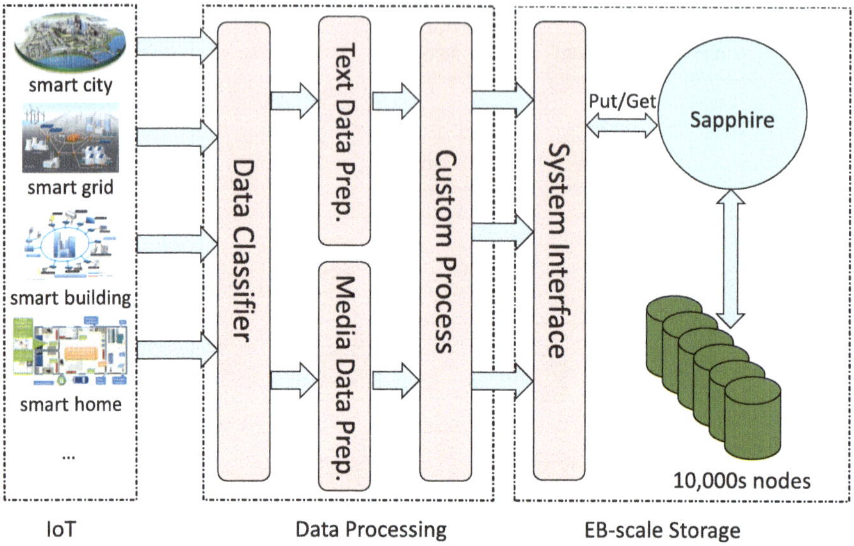

Fig. 4 Sapphire-based IoT system architecture

3.3 System Architecture

As illustrated in Fig. 4, the IoT data comes from smart city [27], smart building [7], smart grid [28], and smart home [4, 5], and we classify data into two types: (1) text data and (2) media data through a *data classifier*. We store the two types of data into a blockchain-based EB-scale storage system through the module of *custom process*. In *Sapphire*, each IoT device is viewed as an OSD. *Sapphire* connects the *System Interface* module via the *Put/Get* APIs. As shown in Fig. 5, *Sapphire* is an EB-scale storage system, in which the hash-based mapping mechanism evenly partitions the key address space into virtual nodes. The virtual nodes (OSDs) are utilized as a means to improve load balancing [29], and they make not only scaling out as data grows, but also cooperative caching [30] and re-distribution become easier. Increasing physical OSDs (nodes) might join/depart, so their virtual nodes might be moved onto them or removed from them seamlessly when scaling out/in. Multiple data replicas are utilized to address the problem of fault tolerance for the failures of storage nodes.

In *Sapphire*, by allocating more virtual nodes per OSD, we reach better namespace locality and load balancing since object identifiers are not evenly distributed. It is not a serious issue from the perspective of space since data structures are typically not so expensive. However, we have to consider a more important problem arising from network bandwidth. In order to keep network connectivity, each node frequently contacts its neighbor nodes to ensure them still alive, and its neighbors are replaced with new neighbor nodes when its old neighbor nodes are not alive any longer. There is a multiplicative increase in network traffic since multiple virtual nodes are running in each super node. However, it is not a serious problem since super nodes are in data

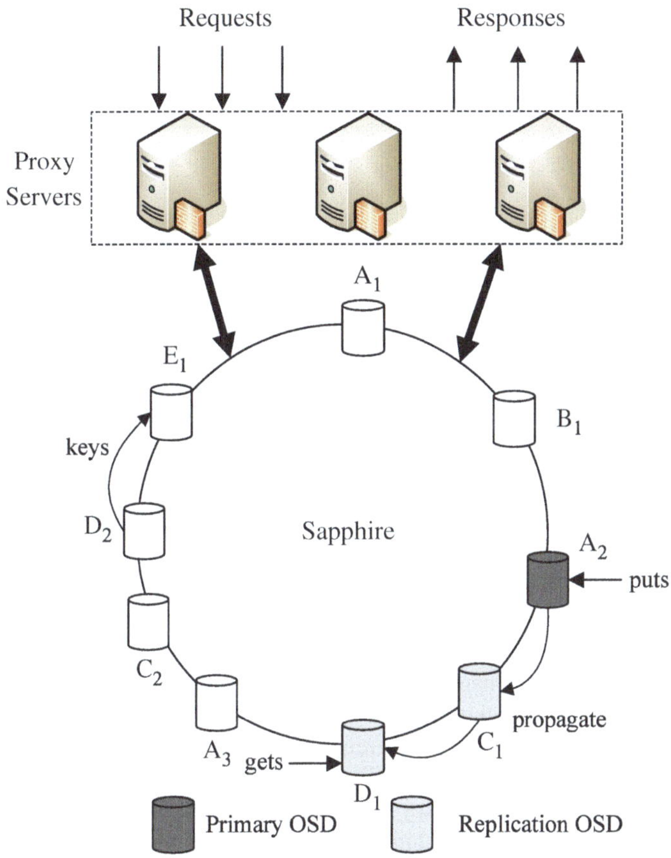

Fig. 5 Sapphire system architecture. Super node *A* has three virtual nodes, regular nodes *C*, and *D* have two virtual nodes, respectively, and light nodes *B* and *E* have only one virtual node, respectively,

centers with enough bandwidth. At the same time, we proposed an efficient linearizable consistency scheme to keep excellent replica consistency among OSDs in our previous work [31].

4 Dynamic Load Balancing

In this section, we present a Location and Type Sensitive (LTS) hashing mechanism for better data analytics in IoT. Due to the LTS hashing mechanism, storage load would not be balanced, so we propose a dynamic load balancing approach to solve this problem.

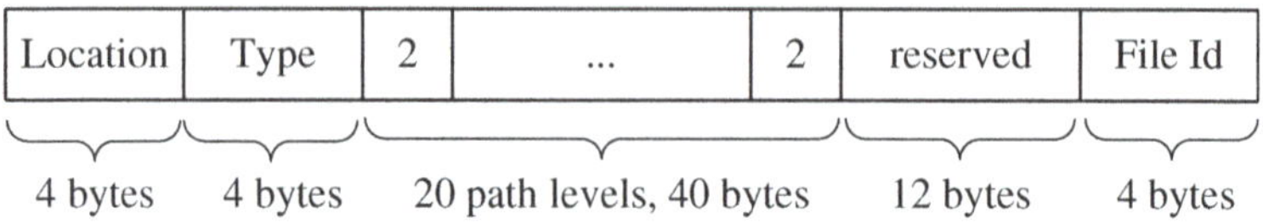

Fig. 6 Location- and type- sensitive hashing

4.1 Location- and Type- Sensitive Hashing

In the LTS hashing scheme, a pathname is directly utilized with a fixed-size key, in which each lookup message contains a key of 64 bytes, as shown in Fig. 6. In order to limit communication overhead without a modification of the given routing mechanism, we employ a compact key by encoding with three fields: *location*, *type* and *pathname*, in Fig. 6. The *location* field is encoded first to place the files with the same location together. The first two fields (i.e., *location* and *type*) are encoded by using four bytes, respectively. In the third filed, each directory is encoded with two bytes. A file name is encoded with the last four bytes, which in theory represent 2^{32} files per directory. Ultimately, the 64-byte key can enable up to many quintillion files with many exabytes of storage in count, so it can excellently satisfy IoT requirements. This key encoding mechanism offers an excellent trade-off between file count and key size, and it can enable naming of new directories and files.

The keys of files are quickly changed to reflect a new path with this key encoding scheme if the files are moved to a different directory. Furthermore, related metadata objects are grouped to preserve in order traversal of file system, e.g., files in the same directory are in a group. Object keys are not uniformly distributed any more in the key space because of the LTS hashing scheme. As shown in Fig. 7, storage load are not balanced. OSDs are in charge of approximately equal ranges of the file key space

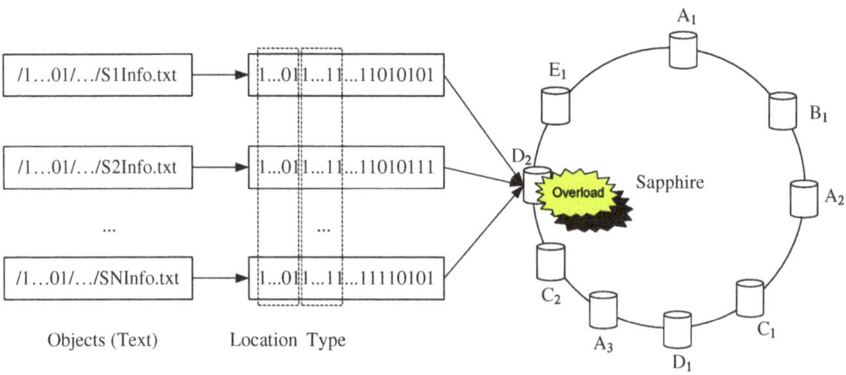

Fig. 7 Load imbalance caused by location- and type- sensitive hashing

in *Sapphire*. However, load balancing is required to limit both the maximum storage space that each OSD has to provision and the data regeneration cost caused upon failures in the worst case.

4.2 Dynamic Load Balancing

Storage load are not balanced anymore because of the LTS hashing mechanism. Each IoT device can periodically contact its neighbors in *Sapphire*. We decide if an IoT device d_i is load balancing only when its storage load L_i meets $1/t \leq L_i/\overline{L} \leq t$ ($t \leq 2$), where $\overline{L}$ is the average storage load of the entire *Sapphire* system. We can say the *Sapphire* system to be in load balancing if the smallest storage load is more than $1/t^2$ of the largest storage load. Suppose that there are a set of m IoT devices $D = \{d_i, i = 1, \dots, m\}$, and there are a set of n virtual nodes $V = \{v_j, j = 1, \dots, n\}$ from m IoT devices, each virtual node v_j has a weight w_j, and each IoT device d_i has a remaining capacity (weight) W_i. w_j means how many files v_j keeps in a key range, and W_i means the difference between the existing weight in the IoT device d_i and the average storage load $\overline{W}$. We can formulate this problem as a 0–1 Multiple Knapsack Problem (MKP) [32]. In other words, it determines how to reassign n virtual nodes to m IoT devices in such a way that is able to minimize the wasted space in the IoT devices. It reads as follows:

$$\text{minimize} \quad z = \sum_{i=1}^{m} s_i \tag{1a}$$

$$\text{s.t.} \quad \sum_{j=1}^{n} w_j x_{ij} + s_i = W_i y_i, i \in M = \{1, \dots, m\} \tag{1b}$$

$$\sum_{i=1}^{m} x_{ij} = 1, j \in N = \{1, \dots, n\} \tag{1c}$$

$$x_{ij} \in \{0, 1\}, y_i \in \{0, 1\}, i \in M, j \in N \tag{1d}$$

$$w_j x_{ij} = \sum_{k=1}^{l} o_{jk}, i \in M, j \in N \tag{1e}$$

where

$$s_i = \text{space left in IoT device} i$$

$$x_{ij} = \begin{cases} 1 & \text{if } v_j \text{ is reassigned to IoT device} i \\ 0 & \text{otherwise} \end{cases}$$

$$y_i = \begin{cases} 1 & \text{if IoT device } i \text{ is used} \\ 0 & \text{otherwise} \end{cases}$$

$$o_{jk} = \text{the } k\text{th object's storage size in } v_j$$

Constraint (1b) makes sure that the total number of files that are assigned to each IoT device is less than the IoT device's capacity. Constraint (1c) ensures that each virtual node is only assigned to a unique IoT device. Constraint (1d) states that it is a 0–1 knapsack problem. Constraint (1e) means there are l objects with distinct sizes, in which o_{jk} is the kth object's storage size in v_j. The *Sapphire* system is different from our previous research DROP [33]: *Sapphire* stores objects with varied sizes, while DROP stores metadata items with fixed sizes.

4.3 Traffic Control

We proved that our dynamic load balancing approach is convergent in [33]. Objects might be moved multiple times during load balancing. *Sapphire* employs pointers to minimize the migration overhead. When the IoT device holds the pointers for longer than their stabilization time, this IoT device retrieves the objects through the pointers. When balancing the storage load, exploring the pointers temporarily hurts data locality. Apart from reducing the overhead of load balancing, the pointers enable writes to succeed even while the target IoT device is at capacity. In addition, the pointers can be employed to divert the objects from heavily loaded IoT devices to lightly loaded ones. However, the IoT device at capacity ultimately sheds some load when balancing the storage load, just causing temporarily additional indirection. We assume that a device B takes some virtual nodes of a device A to take some of A's storage load when A is heavily loaded. A has to transfer its some objects to B. Instead of having A immediately transfer its some objects to B when B obtains some virtual nodes from A, B initially maintains the pointers to A. Later B transfers the pointers to C, and C finally retrieves the actual data from A and removes the pointers.

5 OSD-based Smart Contracts Among IoT Devices

In this section, we present an OSD-based smart contract (*OSC*) mechanism that is used in *Sapphire* as a transaction protocol, in which IoT devices interact with such blockchains.

5.1 IoT Device Coordination

Autonomous device coordination is required in a decentralized IoT solution. Without central third parties, such a decentralized IoT solution grants greater power to the IoT devices' owners to define how the IoT devices interact with each other via the rules of engagement. The decentralized IoT solution recognizes that different IoT devices have varying levels of trust among them depending on operating within constraints imposed by physical proximity and interoperability. In this way, IoT devices are able to engage in autonomous transactions, and they are organized to a decentralized network. In order to achieve it, a smart contract mechanism is equipped to the IoT devices to make contractual agreements with other IoT devices. Besides the security provided by the blockchain protocols [34], the operational security is vital in smart contracts. In a very huge decentralized IoT network including many devices that operate autonomously, the devices are untrusted and some of them might even be malicious. The *Sapphire* system needs to self-organize and achieve consensus-based autonomous coordination to guard against routing or DDOS attacks. In *Sapphire*, super nodes are utilized to guarantee the operational security.

5.2 Smart Contracts

A smart contract executes contract terms, which are recorded on blockchains. It also includes performed obligations and future processes as shown Fig. 8. The emergence of blockchains makes it achievable without involving of third parties. Our OSD-based smart contract (*OSC*) module includes three components: (1) transaction authenticity, (2) data traceability, and (3) system security, as shown in Fig. 9. In the context of blockchains, smart contracts are pre-written logic, in which various

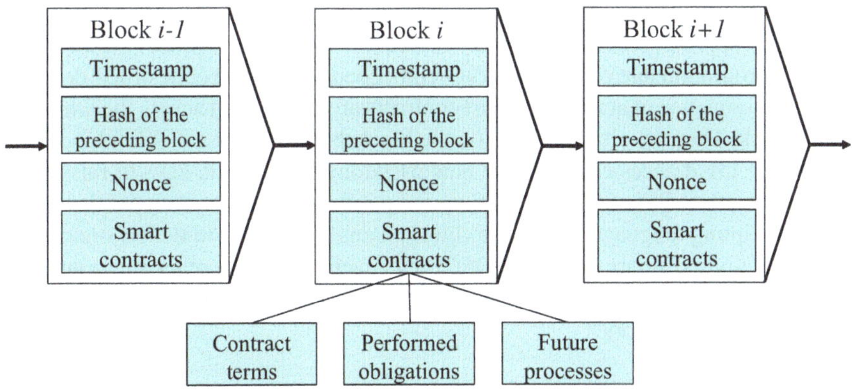

Fig. 8 Blockchain for IoT devices

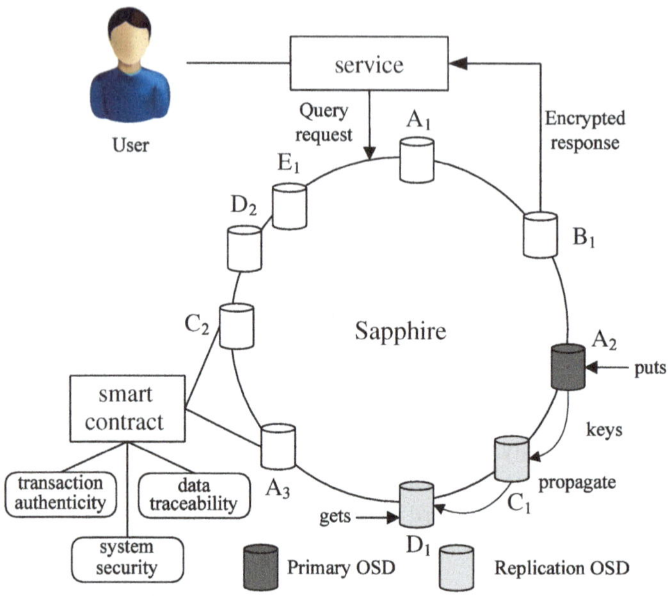

Fig. 9 Smart contracts for IoT devices in the *Sapphire* system

processing tasks are registered as scripts in advance to be executed automatically. Smart contracts are stored and replicated on the distributed storage system *Sapphire* and executed by a network of IoT devices. The component of *transaction authenticity* prevents the same processing task from being executed multiple times and ensures that the execution of a contract among IoT devices can be proved retrospectively. The component of *data traceability* ensures that processing records are traceable in IoT devices. The *system security* component makes sure that the contracts among IoT devices are managed on a blockchain to preserve the records of contracts.

Smart contracts can be used for allocating digital contracts between two IoT devices in *Sapphire*, when the requirements established in the contract are fulfilled. They are programmable contractual tools embedded in software code of IoT devices. The entire sequence of actions in smart contracts are propagated across the *Sapphire* IoT network. They are also recorded on the blockchain, so they are publicly visible. Even if the IoT devices can generate new pseudonymous public keys to raise their anonymities, the values of all transactions are publicly visible for each public key. In *OSC*, a scripting language is added to blockchains and allows to define smart contracts. OSD-based smart contracts (*OSCs*) can even instantiate other sub-contracts. This makes *OSC* possible to implement various forms of contractual agreement in the *Sapphire* IoT network.

6 IoT Data Analytics

Object-based storage is a solution that accesses and manipulates the discrete entities of storage (i.e., objects). Similar to the files, the objects include data, but they are not organized in a hierarchy.

6.1 Use Cases in IoT

The IoT exponentially raises data variety, velocity, and volume. As usual, the burden falls on information technology (IT) to solve the dilemmas of data collection, integration, storage, and analytics, which are caused by the IoT. As shown in Table 1, we demonstrate some popular use cases in the Internet of Things. Current strategies cannot be employed since the data to be captured and explored is even more various. Meanwhile, the IoT use cases are also more diverse. The IoT data coming from a large number of devices is attached to a variety of objects and devices. Many objects may exist at the same level of a flat address space. Both files and objects have metadata that is associated with their data. However, the extended metadata of objects characterizes the objects themselves.

Without the physical locations of objects, the objects can be retrieved through their unique identifiers. Object storage is employed to perform data analytics, where large-scale data sets are scanned for the patterns of different complexities. It efficiently supports online analytical processing (OLAP) tasks, with improvements comparable to the scan-based operations. It also allows many data-intensive appli-

Table 1 Use cases in IoT

Questions	Queries
Which IoT device text files are accessed within three days?	Type = txt, atime < 3 days
How many data files are collected from an IoT device *A* in a region *B* from 05/01/2017 to 08/01/2017?	Dir = /10001, Name = A, When = 05/01/2017:08/01/2017
Which videos will be expired and removed from a building *X* in a region *Y*?	retention time=expired, ctime > 3 months type=video, Dir=/11101/X
What videos are from two cameras *A* and *B* in the same building *C* from 4pm to 5pm in 28/01/2017?	Dir=/11001/C/A, Dir=/11001/C/B, When=4pm:5pm in 28/01/2017, type=video
What's the average degree for all sensors in a region *A* from 09am to 10am in 25/01/2017?	Dir=/10011, files = sensors*.txt, When = 09am:10am in 25/01/2017
How much storage is consumed by the video files in all IoT devices in a building *A*?	Sum size where dir = /10011/A, type = video

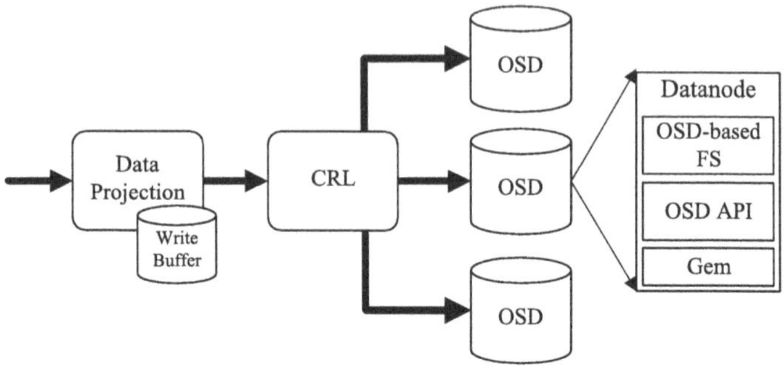

Fig. 10 OSD-based HDFS datanode

cations to explore OSDs with changes to some database (DB) primitives. Many new optimizations are utilized to improve the scheduling knowledge available when data analysis applications are executed in the IoT devices.

6.2 OSD Interfaces with HDFS

In this subsection, we proceed to execute drop-in replacement for HDFS to make it compatible with HDFS, thus existing jobs can be executed without modifications. By implementing the interfaces of HDFS, it can be completed since file system semantics are emulated through object storage, metadata, and indexing. We have to adhere to a rule: map tasks read locally instead of from the network, in a local computation based on *CRL*, which is concurrent regeneration code with local reconstruction [35].

Dispersal raw data falls in contiguous chunks on m nodes when writing, while MapReduce tasks read locally from raw data slices, bypassing erasure code reconstruction when reading. After raw data stream comes, optimized data chunks are calculated and then they are placed in OSDs, as shown in Fig. 10. HDFS is integrated with OSDs by revising the "Default FS" module to communicate with the APIs on the OSDs, as shown in Fig. 10. The integration of HDFS with OSDs does not depend on the Hadoop ecosystem, in which only the Datanodes are modified. The revised "Default FS" (i.e., OSD FS) breaks the object data into smaller chunks since OSDs employ object storage instead of block storage.

6.3 Object Management

As shown in Fig. 11, we manage objects (i.e., records) in a traditional relational DB with user data as BLOBs (Binary Large Objects), e.g., videos and images, which might be stored across a number of blocks. Depending on the requirements for access

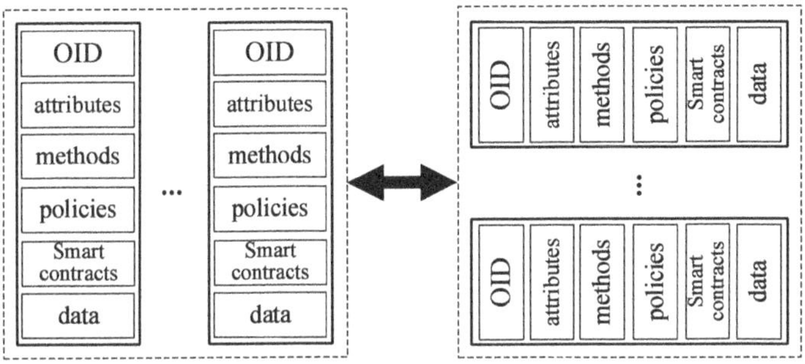

Fig. 11 Managing objects with BLOBs in *Sapphire*

Fig. 12 Object-based data repository in *Sapphire*

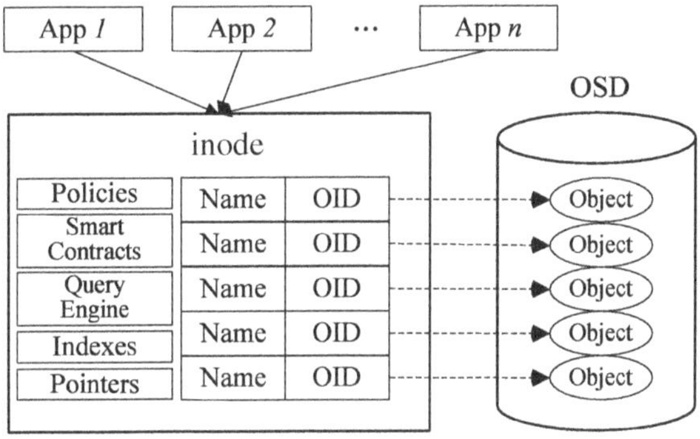

and manipulation, it is necessary to stripe the BLOBs across multiple storage nodes, allowing to concurrently retrieve its content. Files in an object storage system are naturally mapped to objects and stored in the OSDs. An empty object is created via an OSD command: *CREATE*, and the object data is accessed and manipulated via standard OSD *READ* and *WRITE* commands. Furthermore, the OSD instructions offer for optional attributes assigned to objects by specifying useful QoS (Quality of Service) parameters, such as the expected size of the object and prevalent access pattern, e.g., random or sequential.

The *Sapphire* system provides other alternative access methods that are compatible with object storage, starting with those that work within existing specifications. Object-based data repository is illustrated in Fig. 12. In a sequentially ordered file, object ID (OID) is the primary index that is an index whose search key specifies the file's sequential order. Secondary index is another index whose search key is different from the file's sequential order. In other words, the records in the file are not ordered according to secondary index. Pointers are part of the objects, and they are not typical for the objects. The IoT storage system *Sapphire* allows attributes of pointer type, i.e., references. Secondary indexes consists of blocks that usually have the pointers within them. The query engine in an object takes a query plan including join, selection, and aggregation. It executes the plan and returns the result. Policies from upper applications are integrated into OSDs, such as locking, scheduling, and preserving referential integrity.

7 Summary

This chapter explores the use of object-based storage to improve the interactions between storage systems and data analytics applications in IoT. We present a large-scale blockchain-based storage system, called *Sapphire*, for data analytics in the Internet of Things (IoT). We develop an OSD-based smart contract (*OSC*) approach as a transaction protocol, where IoT devices interact with such blockchains in *Sapphire*. Modern data analytics applications approximately explore the typical features of distributed storage systems. However, modern distributed storage systems do not have semantic knowledge for the requirements of data analytics in the Internet of Things. This makes it difficult to design exceptional optimization decisions. In the *Sapphire* system, we use object-based storage interfaces to allow analytics applications to communicate the requirements of storage to the blockchain-based object storage system for the IoT. By complying with standard OSD specifications, *Sapphire* addresses the IoT data at a fine granularity and it allows analytics applications to access and manipulate individual objects and their attributes. As a blockchain-based storage system, *Sapphire* has much richer semantic information for the stored objects to optimize its performance more effectively than other storage systems. With better semantic information, *Sapphire* would better optimize its layout and set aside free space for future operations.

References

1. D. Evans, The internet of things how the next evolution of the internet is changing everything (2011), http://www.cisco.com/web/about/ac79/docs/innov/IoT_IBSG_0411FINAL.pdf
2. K. Rose, S. Eldridge, L. Chapin, The internet of things: an overview understanding the issues and challenges of a more connected world (2015), http://www.internetsociety.org/sites/default/files/ISOC-IoT-Overview-20151022.pdf
3. L. Atzori, A. Iera, G. Morabito, The internet of things: a survey. Comput. Netw. **54**(15), 2787–2805 (2010)
4. C. Dixon, R. Mahajan, S. Agarwal, A. Brush, B.L.S. Saroiu, P. Bahl, An operating system for the home, in *NSDI. USENIX* (2012)
5. J. Vanus, M. Smolon, R. Martinek, J. Koziorek, J. Zidek, P. Bilik, Testing of the voice communication in smart home care. Hum. Centric Comput. Inf. Sci. **5**(15), 1–22 (2015)
6. Z. Fan, P. Kulkarni, S. Gormus, C. Efthymiou, G. Kalogridis, M. Sooriyabandara, Z. Zhu, S. Lambotharan, W.H. Chin, Smart grid communications: overview of research challenges, solutions, and standardization activities. IEEE Commun. Surv. Tutor. **15**(1), 21–38 (2013)
7. F. Zafari, I. Papapanagiotou, K. Christidis, Micro-location for internet of things equipped smart buildings. IEEE Internet Things J. **3**(1), 96–112 (2016)

8. T. Hardjono, N. Smith, Cloud-based commissioning of constrained devices using permissioned blockchains, in *Proceedings of the International Workshop on IoT Privacy, Trust, and Security* (2016), pp. 29–36
9. K. Christidis, M. Devetsiokiotis, Blockchains and smart contracts for the internet of things. IEEE Access **4**, 2292–2303 (2016)
10. R. Pass, L. Seeman, A. Shelat, Analysis of the blockchain protocol in asynchronous networks. IACR ePrint (2016)
11. G. Wood, Ethereum: a secure decentralized transaction ledger, http://gavwood.com/paper.pdf
12. M. Mesnier, G.R. Ganger, E. Riedel, Object-based storage. IEEE Commun. Mag. **41**(8), 84–90 (2003)
13. Q. Xu, K.M.M. Aung, Y. Zhu, K.L. Yong, A large-scale object-based active storage platform for data analytics in the internet of things, in *The 9th International Conference on Multimedia and Ubiquitous Engineering (MUE)* (2015), pp. 405–413
14. Q. Xu, K.M.M. Aung, Y. Zhu et al., Building a large-scale object-based active storage platform for data analytics in the internet of things. J. Supercomput. **72**, 2796–2814 (2016)
15. G.A. Gibson, R.V. Meter, Network attached storage architecture. Commun. ACM **43**(11), 37–45 (2000)
16. N. Szabo, Formalizing and securing relationships on public networks. First Monday **2**(9) (1997)
17. L. Luu, D.H. Chu, H. Olickel, P. Saxena, A. Hobor, Making smart contracts smarter, in *ACM CCS* (2016)
18. J. Wang, P. Shang, J. Yin, Draw: a new data-grouping-aware data placement scheme for data intensive applications with interest locality, in *Cloud Computing for Data-Intensive Applications* (Springer, 2014), pp. 149–174
19. E. Riedel, G.A. Gibson, C. Faloutsos, Active storage for large-scale data mining and multimedia, in *VLDB* (1998), pp. 62–73
20. A. Acharya, M. Uysal, J.H. Saltz, Active disks: programming model, algorithms and evaluation, in *ASPLOS* (1998), pp. 81–91
21. K. Keeton, D.A. Patterson, J.M. Hellerstein, A case for intelligent disks (idisks). SIGMOD Rec. **27**(3), 42–52 (1998)
22. L. Huston, R. Sukthankar, R. Wickremesinghe, M. Satyanarayanan, G.R. Ganger, E. Riedel, A. Ailamaki, Diamond: a storage architecture for early discard in interactive search, in *FAST* (2004), pp. 73–86
23. S.W. Son, S. Lang, P. Carns, R. Ross, R. Thakur, B. Ozisikyilmaz, P. Kumar, W.K. Liao, A. Choudhary, Enabling active storage on parallel I/O software stacks, in *MSST* (2010), pp. 1–12
24. Q. Xu, H.T. Shen, Z. Chen, B. Cui, X. Zhou, Y. Dai, Hybrid retrieval mechanisms in vehicle-based P2P networks, in *Proceedings of the International Conference on Computational Science (ICCS'09)*. Lecture Notes in Computer Science, vol. 5544 (Springer, Berlin, 2009), pp. 303–314
25. K. Shvachko, H. Kuang, S. Radia, R. Chansler, The hadoop distributed file system, in *MSST* (2010), pp. 1–10
26. Q. Xu, Y. Dai, B. Cui, A HIT-based semantic search approach in unstructured P2P systems. Acta Sci. Nat. Univ. Pekin. **46**(1), 17–29 (2010)
27. Y. Li, W. Dai, Z. Ming, M. Qiu, Privacy protection for preventing data over-collection in smart city. IEEE Trans. Comput. **65**(5), 1339–1350 (2016)
28. N. Boumkheld, M. Ghogho, M.E. Koutbi, Energy consumption scheduling in a smart grid including renewable energy. J. Inf. Proces. Syst. **11**(1), 116–124 (2015)
29. I. Stoica, R. Morris, D.R. Karger, M.F. Kaashoek, H. Balakrishnan, Chord: a scalable peer-to-peer lookup service for internet applications, in *SIGCOMM* (2001), pp. 149–160
30. Q. Xu, H.T. Shen, Z. Chen, B. Cui, X. Zhou, Y. Dai, Hybrid information retrieval policies based on cooperative cache in mobile P2P networks. Front. Comput. Sci. China **3**(3), 381–395 (2009)
31. Q. Xu, R.V. Arumugam, K.L. Yong, S. Mahadevan, Efficient and scalable metadata management in EB-scale file systems. IEEE Trans. Parallel Distrib. Syst. **25**(11), 2840–2850 (2014)

32. C. Chekuri, S. Khanna, A polynomial time approximation scheme for the multiple knapsack problem. SIAM J. Comput. **35**(3), 713–728 (2005)
33. Q. Xu, R.V. Arumugam, K.L. Yong, S. Mahadevan, DROP: facilitating distributed metadata management in EB-scale storage systems, in *MSST* (2013), pp. 1–10
34. A. Kosba, A. Miller, E. Shi, Z. Wen, C. Papamanthou, Hawk: the blockchain model of cryptography and privacy-preserving smart contracts, in *IEEE Symposium on Security and Privacy (S&P)* (2016), pp. 839–858
35. Q. Xu, W. Xi, K.L. Yong, C. Jin, Concurrent regeneration code with local reconstruction in distributed storage systems, in *The 9th International Conference on Multimedia and Ubiquitous Engineering (MUE)* (2015), pp. 415–422

Ensuring Quality of Service in the Internet of Things

Giacomo Tanganelli, Carlo Vallati and Enzo Mingozzi

Abstract The Internet of Things is expected to radically reshape many processes in a broad range of domains, from personal to industrial. In many of these heterogeneous scenarios, IoT systems will need to guarantee required levels of reliability and latency in order to provide high-quality services to end users. Quality of service support in IoT system will demand for explicit support at different levels. At the network level, on the one hand, specific technical communication standards will be necessary to ensure timed and reliable data delivery. At the application level, instead, dedicated support from application protocols and design of novel resource allocation algorithms will be mandatory to cope with concurrent access and implement proper management of resources. In this chapter, an overview of the current solutions for ensuring QoS in IoT systems is provided. Specifically, we first survey the current approaches at the network level through a summary of all the mechanisms included in the main communication standards for IoT. We then deliver an analysis of the current solutions available in IoT protocols and platforms to enforce QoS at the application level.

Keywords Quality of service · Network protocols · IoT platforms

1 Introduction

The continuously increasing sensing, actuating, computing, and communication capabilities embedded into everyday objects around us are turning the earlier vision of the Internet of Things into real. Many solutions are commercially available today that exploit networked "smart" objects to provide end users with advanced services connected to the physical world. Such solutions are, however, often vertical, isolated systems based on ad hoc HW/SW realizations which are not able to cooperate with each other to share common smart object capabilities for efficiency and scalability.

G. Tanganelli · C. Vallati (✉) · E. Mingozzi
University of Pisa, Pisa, Italy
e-mail: c.vallait@iet.unipi.it

R.R. Yager and J. Pascual Espada (eds.), *New Advances in the Internet of Things*,
Studies in Computational Intelligence 715, DOI 10.1007/978-3-319-58190-3_9

Isolation, however, is not the only drawback; from a software developer perspective, the lack of a common software fabric allowing to interact with smart objects entails great limitations on software portability and maintenance. To overcome such limitations, a horizontal approach to develop and deploy IoT applications is by far more appropriate and desirable. To this aim, middleware platforms exposing standard interfaces to access smart objects are currently under definition in order to allow an easy integration of heterogeneous (and already existing) systems, and to facilitate the development of application logic based on a converged infrastructure.

IoT systems are expected to be employed in a large number of use cases for heterogeneous applications. For some of them, such as smart industrial or smart grid systems, explicit support for quality of service (QoS) is mandatory. In such systems, in fact, transmission failures and exceeding the required delay bounds are intolerable as they usually cause system instability and outages that might lead to economic and material losses and, ultimately, a threat to human safety. Offering an effective fine-grained QoS support in IoT systems requires specific functionalities to be implemented at different levels. As in traditional systems, QoS support at the network level is mandatory in order to guarantee fulfillment of QoS requirements in the communication between IoT devices and applications. In addition, IoT systems will require support at the application level, in both application protocols and platforms, in order to cope with concurrent applications and implement efficient management of shared resources.

In this chapter, we offer an overview of the QoS support available in the current IoT technologies. The analysis covers the aspects related to network and application levels. The content of the chapter is structured as follows: first, an analysis of the different use cases in which QoS is mandatory is offered, and then an overview of the QoS support available at the network level and at the application level is provided. Eventually, an overview of future research directions related to QoS support in IoT concludes the chapter.

2 IoT Use Cases with QoS Requirements

QoS support in IoT systems is mandatory in a large number of use cases. Each scenario, however, will be characterized by a different set of QoS requirements that can vary noticeably among each other. In the following, the main IoT use cases that require QoS support to ensure proper operation of systems are presented along with a short characterization of their main requirements.

2.1 Smart Manufacturing

The rapid evolution of IoT technologies has recently captured the attention of industrial companies that expect IoT systems to introduce a breakthrough to enhance the efficiency of the manufacturing process. This emerging use case, referred in the literature as Industrial IoT or IIoT for short, represents a significant challenge for IoT deployments. In a typical IIoT system, sensors and actuators are deployed in a dedicated network inside a factory plant to collect specific data and to assist and control the production process. Although the architecture of an IIoT system is not different from a standard IoT system, the requirements that many industrial manufacturing processes demand represent the main challenge. In fact, stringent QoS requirements in terms of resiliency, reliability, and latency are mandatory to ensure proper implementation of manufacture automation [1].

An example of smart manufacturing IoT applications is a closed-loop control for non-critical processes. In this case, the application requires that the telemetry data and the control commands, from sensors and to actuators deployed in the assembly line, respectively, be strictly delivered with latencies by **tens of milliseconds** [2]. When a higher delay is experienced, the whole system enters into an emergency shutdown state, which might cause substantial financial repercussions.

Emergency signals produced by sensors are characterized by even more stringent requirements, as they must be dispatched to a powerful central controller with the lowest delay possible. Furthermore, in this case, the reliability of communication is extremely important since a packet loss may result in products with defects. In this case, enforcement of QoS requirements is more challenging since it is not only limited to time-related parameters, but involves different aspects, such as reliability, which can be achieved by introducing a certain level of redundancy in the system architecture, in order to promptly react to system failures.

In order to deal with such strict requirements, different wireless communication standards have been specifically designed for industrial applications. Among them, it is worth to mention the WirelessHART and 6TiSCH protocols.

2.2 *Smart Grid*

The introduction of renewable energy sources, potentially distributed over a wide area, is increasing the complexity of the electric grid system. The creation of local bidirectional energy flows introduced the need for an efficient and flexible management infrastructure that triggered the evolution toward a smart grid system, in which a new communication infrastructure is exploited [3]. Such infrastructure aims at enabling real-time monitoring and coordination with the use of smart devices that employ machine-to-machine (M2M) communications throughout the overall electric network [4].

Smart grid networks are characterized by heterogeneous topologies composed by different functional blocks (e.g. generation, transmission, and distribution.) deployed along the chain from the supplier to the consumer. Although a single representative topology cannot be derived, three different types of networks are usually involved: Wide area networks (WANs), field area networks (FANs), and home area networks (HANs). WANs are used to connect power plants to local

Table 1 IEEE 1646 communication delivery time performance requirements for electric power substation automation

Information types	Internal to substation	External to substation
Protection information	4 ms	8–12 ms
Monitoring and control information	16 ms	1 s
Operations and maintenance information	1 s	10 s
Text strings	2 s	10 s
Processed data files	10 s	30 s
Program files	60 s	10 min
Image files	10 s	60 s
Audio and video data streams	1 s	1 s

substations (substations convert high-voltage electricity to low-voltage for distribution purposes). FANs, instead, are used between close substations to monitor and control the electric behavior in a neighborhood, performing small independent adjustments depending on the required load. Finally, HANs are exploited by smart devices at customer premises to implement new functionalities, such as automatic metering, which leverages on information provided by the electric supplier.

Network heterogeneity leads to different communication requirements that must be guaranteed for safety reasons. To this aim, a smart grid system has to include QoS support to ensure reliable and timely communication. Above all the requirements, timing is the most critical, as some type of information are useful only when delivered within a predefined deadline. When a deadline is missed, damage might incur in the grid system. The IEEE 1646 standard [5] formally defines the communication latencies required in electric grid systems. These requirements are summarized in Table 1 according to different information types and according to the different communication end points.

The implementation of smart grid networks is performed using different communication technologies [6]. Considering the QoS requirements similar to industrial applications, the same wireless technologies adopted for smart manufacturing are often exploited in FANs and HANs. The realization of WANs, instead, is performed through long-range cellular networks, which embed direct QoS support by design. Smart grid networks can adopt also the powerline communication (PLC) technology that uses the existing power line to transmit data signal. However, its harsh and noisy channel environment makes hard the implementation of any QoS support, which restricts the applications of the PLC technology.

2.3 eHealth

Latest technology developments will enable new frontiers for the IoT. Among them, smart health, or eHealth, is a promising use case that is expected to improve significantly the quality of our lives providing new healthcare services such as remote patient monitoring. Also in this context, QoS guarantees are a key requirement [7]. In remote health monitoring, for example, through a body sensor network (BSN), the collected data have different relevance, e.g., heart activity data are more important than data on the body temperature. For this reason, collection and delivery of data must be prioritized accordingly through different QoS levels. In addition, data priority can dynamically change over time depending on the sensor value, e.g., glucose data priority increases when samples indicate hypo- or hyper-glycaemia.

To this aim, the IEEE 1073 working group has defined QoS requirements for several health applications. For instance, the application that is characterized by the most stringent QoS requirements is electrocardiogram (ECG) monitoring. Such application requires sending bursts of 4 kbit/s of data that must be delivered within a maximum delay of 500 ms for each electrode [8].

The implementation of smart health monitoring systems exploits Bluetooth as communication technology [9]. The latter embeds support for short-range communication with controlled latency.

3 QoS Support in IoT Networks

Timely and reliable communication is of paramount importance in providing QoS assurances for IoT applications. To this aim, several communication standards have been defined for both wireless and wired networks. Future IoT deployments are expected to heavily leverage wireless communications, as they ensure rapid deployment, broad coverage, and low cost. For this reason, QoS support will be crucial in future wireless standards for IoT. In this section, an overview of the major wireless standards designed for IoT is offered. The presentation covers only the standards that include QoS support by design.

3.1 WirelessHART

WirelessHART has been developed in 2008 as an extension of the HART protocol used in industrial plants for wired communication since ‘80s [10]. It provides the same interface of its wired counterpart to allow easy integration, but the network, data-link, and physical layers have been re-designed in order to provide the same assurances that are available in the HART protocol. WirelessHART operates in the 2.4 GHz ISM radio band by exploiting IEEE 802.15.4 compatible DSSS radios with channel hopping on a packet-by-packet basis. The network is organized as a multi-hop mesh network, in which each device can relay packets from neighbors to achieve multi-hop communication, which guarantees reliable message delivery exploiting the alternative paths available in case of failures. Both unicast and multicast traffic are supported.

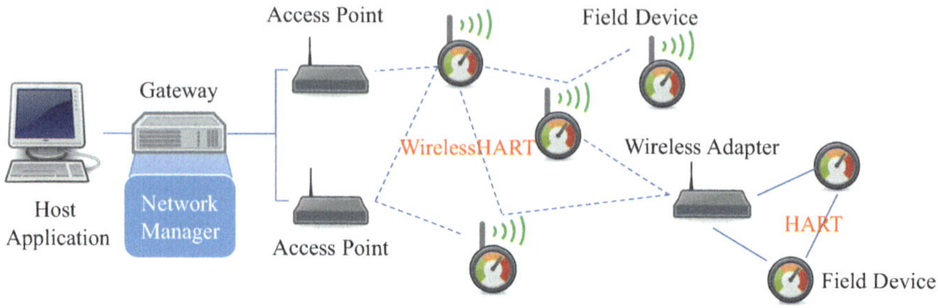

Fig. 1 WirelessHART devices

The WirelessHART protocol employs time division multiple access (TDMA) as MAC protocol implemented through precise time synchronization among devices. With TDMA, time is divided into timeslots, whose duration is sufficient to send or receive one packet per channel and the corresponding acknowledgment to ensure reliable transmission.

The protocol identifies different types of devices that cooperate to form the overall network, see Fig. 1:

- *Field devices*: sensing and actuating devices that generate or receive data. When required, field devices relay traffic from/to neighbors.
- *Routers*: devices that only relay packets through the wireless network.
- *Adapters*: devices that bridge wired HART field devices with the WirelessHART network.
- *Access points*: devices that connect field devices to the gateways.
- *Gateways*: devices that connect the WirelessHART network to IP networks.
- *Network manager*: a device that handles the configuration of the whole network.

Core component of the network is the network manager that is responsible of configuring the routing and allocating the transmission opportunities. To this aim, it dynamically builds a routing graph, which is continuously updated based on the status information provided by the devices of the network. The communication between field devices is always based on commands and responses. WirelessHART, in fact, has been designed with a particular focus on field device communications. However, external applications can also use the gateway to schedule traffic for field devices. Every time a field device wants to interact with another field device, and it interrogates the network manager providing the desired communication requirements. The network manager collects all requirements and computes routing and scheduling, which is then mapped into timeslot allocations, sent to each device. Considering that communication among devices often takes place periodically, timeslots are scheduled into periodic superframes of variable length.

The centralized network management and the TDMA scheduling allow WirelessHART to enforce strict QoS requirements for the communication between field devices. To this aim, communications are classified into four different priority levels to prioritize message queuing and delivery: *command*, *process data*, *normal,* and *alarm*. Command messages have the highest priority and are always propagated through the network, thus allowing the network manager to keep the network operational. Alarm messages, despite the name, are low priority messages that can tolerate delay since they are always time-stamped. Finally, all other traffic flows through the network as buffer space and bandwidth allows. Within this traffic, process data has the highest priority. These different levels of priority are exploited whenever retransmissions occur. If a high priority packet needs to be retransmitted, it can be scheduled on a timeslot, allocated to the same field device, which was initially dedicated to a different type of traffic with lower priority. Consequently, a lower priority message may be delayed to allow the higher priority message to be transmitted.

Routing and scheduling policies implemented by the network manager, although crucial for QoS, are not specified by the WirelessHART specifications. For this reason, different algorithms have been proposed in the literature. In [11], authors propose a WirelessHART tailored optimal branch-and-bound algorithm and a pseudo-polynomial heuristic algorithm to schedule dynamic real-time traffic with predefined deadlines. A similar problem is addressed in [12] where authors design an algorithm to minimize the data evacuation time in binary trees, which are common in WirelessHART network topologies, i.e., all communications must traverse the gateway.

3.2 6TiSCH

The 6TiSCH protocol, currently under standardization within IETF, represents the most promising recent standardization effort to foster the new breed of time-sensitive networks in which time-sensitive traffic is prominent. The main goal of this standard is to define a network architecture that supports critical traffic characterized by high sensitivity to both jitter and latency and requires minimal loss. These requirements are typical in command and control networks and are expected, for example, in smart automation applications.

The 6TiSCH protocol stack [13] is based on existing standards, as shown in Fig. 2. Specifically, the MAC protocol is based on the IEEE802.15.4e time slotted channel hopping (TSCH) protocol designed to offer deterministic communication for industrial-type applications. IEEE 802.15.4e TSCH specifies a time slotted access in which nodes transmit/receive on scheduled transmission/reception opportunities, called *cells*, to eliminate collisions among competing nodes thus increasing network throughput. Multi-channel and channel hopping are also exploited to mitigate the effects of interference. On top of the MAC layer, the 6top layer [14] is currently under standardization as a logical link control (LLC) sublayer that abstracts the underneath TSCH network as an IP link on which IPv6 packets, compressed following the 6LoWPAN specification, can be delivered. Multi-hop

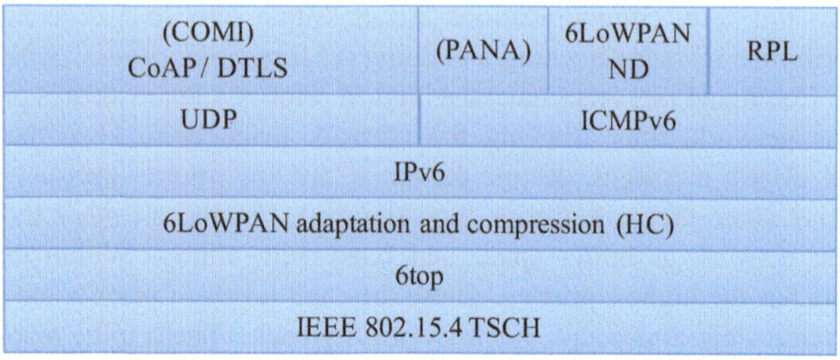

Fig. 2 6TiSCH protocol stack

packet delivery is achieved by means of the RPL routing protocol, the standard de facto protocol for low-power and lossy networks. The routing protocol is responsible for creating an acyclic graph to define all the routes between nodes, thus defining *the path a packet must traverse to be forwarded from one node to another.*

Core of the 6TiSCH architecture is the 6top layer, which is responsible for managing the schedule of the TSCH cells, allocating transmitting and receiving opportunities for all the nodes. To this aim, 6TiSCH defines two different scheduler computation models: one distributed and another centralized. In the former case, the network employs a distributed scheduling mechanism in which each node is responsible for negotiating with its neighbor single transmission opportunities. Routing is performed according to the information provided by the RPL protocol. In the centralized mode, instead, a centralized entity, called path computational element (PCE), is employed to configure the network. In this case, the PCE is responsible for performing the routing, overriding the path derived by the RPL protocol, and for implementing a centralized scheduling policy that allocates the cells to all the nodes in the network. In order to differentiate cells scheduled by the PCE from the cells scheduled locally, two types of cells are defined: hard and soft cells, respectively. Hard cells can be allocated only by a centralized entity and cannot be reallocated dynamically. Soft cells, instead, can be reallocated by 6top dynamically.

The 6TiSCH protocol defines four different paradigms to schedule cells: *Static Scheduling*, *neighbor-to-neighbor scheduling*, *hop-by-hop scheduling* and *remote monitoring and Scheduling Management.* In the static scheduling, a fixed schedule is configured for the whole network. All cells are shared, and nodes contend for cell access in a slotted aloha manner. Such kind of scheduler cannot provide any deterministic guarantee. In the neighbor-to-neighbor scheduling, cells are dynamically allocated by the 6top layer, according to a specific scheduling function (SF). A group of cells allocated between two peers is called a bundle and is exploited by the 6top layer to implement the IP link abstraction with a required bandwidth. The bandwidth of the link is proportional to the number of cells in the bundle. The 6top layer may dynamically adjust the size of a bundle to react to variations in the bandwidth demand. In the hop-by-hop scheduling, a distributed hop-by-hop reservation is performed to reserve a dedicated track between two nodes, e.g., for a specific traffic flow. The reservation is performed through an end-to-end signaling protocol, e.g., RSVP, that allows the 6top layer of each node to allocate the required soft cells for the flow. Contrary to the other modes, in the remote monitoring and scheduling management mode, a centralized scheduling is employed. In order to enable remote management and monitoring of nodes, every node must expose a REST interface to allow an external entity to interact with the 6top layer of the node.

3.3 6TiSCH QoS

The 6top layer is responsible for guaranteeing the QoS requirements specified by the upper layer. QoS enforcement, however, is achieved by the 6top layer not only through the configuration of packet forwarding and the scheduling of cells, but also through the queuing management at each node. To this aim, 6top implements multiple outgoing queues on each node that are exploited to prioritize outgoing packets. Packet classification is performed exploiting their additional properties, such as the next hop neighbor (DestAddr), the TrackId, or the Priority. To achieve this behavior, different IETF working groups (e.g., the Detnet WG) are currently defining a common protocol to prioritize traffic. Among them, the specifications based on the DiffServ classification are the most promising. For example, the Differentiated Service Class Recommendations for LLN Traffic [15] defines different classes of traffic that are common in low-power and lossy networks (LLN). Each class presents different unique characteristics. Every class is mapped to a different DSCP in order to define a deterministic per hop behavior (PHB). Finally, it is worth to note that this classification can be also mapped to the "Industrial Routing Requirements in Low-Power and Lossy Networks" RFC 5673 [2]. The classes, and their corresponding mapping, are summarized in Table 2.

The scheduling policy adopted in 6TiSCH networks is a core component in enforcing the QoS. To this aim, different solutions have been proposed in the literature with different strategies and objectives. Among them, it is worth to mention [16], in which a distributed algorithm is proposed to allocate cells dynamically according to the actual traffic load. Such simple approach is suitable especially for very constrained devices since it exploits only a minimum set of local information. In [17], a similar approach is proposed to achieve a decentralized traffic aware scheduling aimed at minimizing end-to-end latency. This approach, instead of using only local information, exploits neighbor-to-neighbor signaling to schedule the cells along the path between source and destination nodes.

3.4 Bluetooth

The Bluetooth protocol has been standardized more than 15 years ago for wireless personal communications. Its recent revision, Bluetooth Low Energy (BLE), is a

Table 2 LLN traffic classes

Traffic class	Traffic characteristics	Tolerance to			DSCP	RFC 5673 class
		Loss	Delay	Jitter		
Alerts	**Size** = small **Rate** = 1-few **Short** flow **Burst** = none to somewhat	Low	Low	N/A	CS5	2, 3
Control	**Size** = variable, typically small **Rate** = few **Short** flow **Burst** = none to somewhat	High	Low	High	CS5	2, 3
Deterministic control	**Size** = variable, typically small **Rate** = few **Short** flow **Burst** = none to somewhat	Low	Very low	Very low	EF	1
Video	**Size** = big **Rate** = variable **Long** flow **Burst** = non-bursty	Low	Low–Medium	Low	CS3	N/A
Query-based	**Size** = variable **Rate** = variable **Short** flow **Burst** = bursty	Low	Medium	High	AF21 AF22 AF23	4
Periodic	**Size** = variable **Rate** = constant **Long** flow **Burst** = bursty	High	Medium–High	High	AF11 AF12 AF13	5

low-power radio technology with the highest potential for IoT to connect a large variety of personal devices, such as wearable sensors, tablets, smartphones, or smart gadgets. In order to ease the integration of BLE nodes into existing IoT systems, the most recent Bluetooth specification introduced the Internet protocol support profile (IPSP) to offer Internet connectivity to BLE smart sensors through a BLE gateway [18]. In addition to this, IETF recently standardized 6LoWPAN for BLE nodes to enable the delivery of IPv6 packets [19]. The latter, in particular, opened new application areas for personal IoT systems, enabling the exploitation of smartphones as gateways to connect surrounding BLE-enabled sensors [20].

Bluetooth allows the creation of small networks, called piconets, in which there is a master device (the piconet initiator) and up to seven slave devices. However, multiple piconets are allowed to be connected to form a scatternet through devices that operate in multiple piconets at the same time. Time is divided into slots whose access is controlled by the master. The master periodically polls the slaves to request for the amount of data to be transmitted. If a polled slave does not have any data to send, it replies with an empty packet (NULL packet). This feedback is then used by the master to perform slot scheduling according to the slave's buffer status.

In Bluetooth, two different types of physical links can be exploited: synchronous connection-oriented (SCO) and asynchronous connection-less (ACL) links. On top of the physical link, each device implements a link manager protocol (LMP) for link management and QoS enforcement. In addition to LMP, each device implements the logical link control and adaptation protocol (L2CAP), which communicates with the L2CAP modules of other devices for QoS negotiation.

When a SCO link between the master and a slave is established, at least one slot is scheduled for SCO communication. In this slot(s), the master polls the slave periodically to ensure constant bandwidth allocation. This type of service is intended for applications that require fixed or bounded delay, e.g., voice transmission. SCO traffic has precedence over other types of traffic in order to ensure proper network functioning. When an ACL link is established, one or more ACL slots are scheduled for ACL communication; however, the master polls the slave sporadically without time guarantees. Consequently, the bandwidth of an ACL link depends on how often the master polls the slave, which is negotiated during link setup. Setting a minimum polling frequency for each of the slaves ensures minimum bandwidth.

3.5 *Bluetooth QoS*

In a Bluetooth network, the master device is responsible for the enforcement of the negotiated QoS through slot allocation. The scheduling policy is not defined; however, it is specified that it must enforce the QoS parameters (reported in Table 3) that are negotiated during link setup by the L2CAP and LMP protocols. Although protocol specification suggests the use of a Round Robin scheduling policy, several alternatives have been defined in the literature. Among them, an

Table 3 Bluetooth QoS parameters

Name	Component	Description
Token rate	L2CAP	The nominal normal rate
Token bucket size	L2CAP	The maximum burst size
Peak bandwidth	L2CAP	The maximum bandwidth
Latency	L2CAP	The maximum delay between higher layers to physical
Delay variation	L2CAP	The maximum jitter
Flush time out	L2CAP	The maximum retransmit time-out before delete unsuccessful packets
Tpoll	LMP	The maximum interval between consecutive polls

example is the work presented in [21], in which a strategy to implement both polling and scheduling is proposed to reduce intra-piconet packet delay and maximize the overall throughput. The proposed approach exploits the queue status and delay information to adapt the scheduling and the polling interval to the actual traffic, improving link utilization and resource management efficiency.

Although the Bluetooth standard is specifically designed for personal connections, its native QoS support has attracted also the research community to evaluate its usage for industrial applications. In [11], for instance, the authors study the possible application of Bluetooth for industrial applications that require reliable and low-latency communications. The authors specifically investigate a real-time scheduling algorithm for both periodic and aperiodic traffic to reduce the packet loss. The performance of the algorithm with Bluetooth is evaluated against an implementation exploiting IEEE 802.15.4 as wireless standard, demonstrating how Bluetooth can ensure lower communication latency.

4 QoS Support for IoT Applications

In order to define a common interface to interact with IoT devices, different application-level protocols have been standardized. Among the others, the constrained application protocol [22], the data distribution service protocol [23], and the MQ telemetry transport (MQTT) [24] define a common interface exposed by IoT devices to offer their functionalities to applications. Enforcement of the QoS requires explicit support from the application protocols, for example, to specify QoS requirements for the interaction with the IoT device, e.g., the frequency with which data is received from a sensor. Considering also that multiple applications can access the same IoT device concurrently, the application protocol might include also support for specifying the priority of a request, in order to prioritize important information, e.g., an alarm.

A standard IoT application protocol, however, is not sufficient to enable the creation of large-scale IoT systems. Direct interaction between applications and IoT devices is feasible only on a small-scale where the number of devices and concurrent applications is limited. Design and deployment of IoT platforms is widely considered as an essential element for the implementation of large-scale IoT systems. Their implementation requires an architecture to support complex functionalities such as discovery, and to guarantee scalability in deployments that can involve thousands of devices and can span over different domains.

Implementation of QoS support in large-scale systems requires explicit support from the platform to implement proper management of resources. The latter can affect significantly the resulting QoS, as it is responsible for managing the requests from applications, assigning the actual IoT device for retrieving data or triggering an action. To this aim, a QoS-aware IoT platform is required to ensure that devices are not overloaded. Eventually, the platform is also responsible for interacting with the network to configure the network QoS based on the allocation.

Different IoT platforms are currently available; however, only a few examples include explicit QoS support. Although QoS features can be introduced (and will be) in current IoT platforms, their implementation is still a research open issue. To this aim, only experimental implementations are available in platforms developed as part of research projects.

In this section, an overview of the QoS support provided at the application level is provided. The first part illustrates DSS, CoAP, and MQTT, three popular IoT application protocols that provide a minimal QoS support. The second part, instead, presents three examples of IoT platforms that include experimental QoS support: the oneM2M platform, which defines interfaces for QoS configuration at the network layer; the BETaaS platform, which implements actual support for the management of application requests; and the IoT@Work platform, which includes a minimal support for network layer QoS configuration.

4.1 Data Distribution Service (DDS)

DDS is a novel publish/subscribe protocol for IoT applications based on topics [23]. DDS can be exploited by both sensors and applications to exchange information of interests. Topics are uniquely identified by their name, whose structure also allows the definition of patterns.

Each Publisher has at least one DataWriter module, while each subscriber has at least one DataReader module. Applications that want to publish send the data to the DataWriter that is responsible for the delivery of information to all the interested DataReaders. Each DataReader receives the information and sends the data to the interested applications. The specific implementation of the discovery operations are left out of the DDS specifications and partially covered in other documents, e.g., DDSI-RTPS [25].

The protocol is agnostic to the underlying network protocols. In this respect, the DDSI-RTPS [25] specifications suggest the adoption of UDP as it does not introduce any delay (such as TCP).

4.2 DDS QoS

DDS is one of the few IoT application protocols that includes explicit QoS negotiation by design. To this aim, each DataReader is associated to a DataWriter only if the topic matches and the QoS guarantees offered by the Writer are satisfied by the requirements specified by the Reader. DDS defines different QoS policies from different point of views; however, it is important to remark that each QoS profile has a specific per-topic-behavior, customizable with the parameters presented in the following.

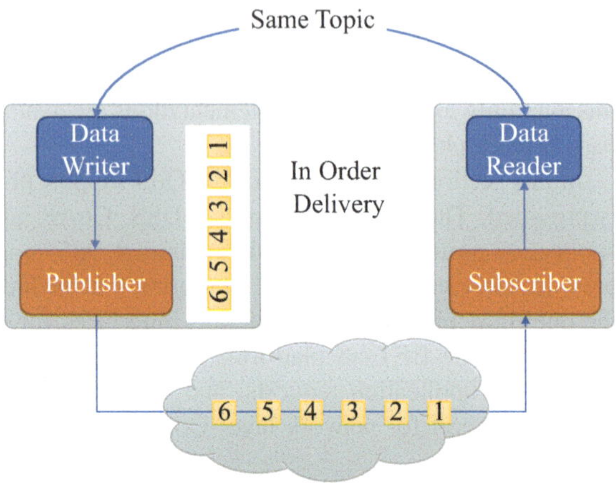

Fig. 3 DDS exchange

DDS defines two different *delivery methods*: RELIABLE and BEST_EFFORT. The former is based on a simple stop-and-wait acknowledgment mechanism, in which the number of outstanding transactions is bounded to one, and a time-out is associated to each message. The BEST_EFFORT, instead, does not provide any guarantee but the order of message delivery. This functionality is implemented through a time stamp that is exploited by the DataReader to reconstruct the correct stream of packets.

In order to guarantee *timely delivery*, DDS provides three different policies: DEADLINE, LATENCY_BUDGET, and TRANSPORT_PRIORITY. When cDEADLINE is used, applications can specify the *maximum* inter-arrival time for the data. This mode can be used when the application wants a periodic stream of data with a fixed period. LATENCY_BUDGET is defined for delay-tolerant applications, which, however, can specify a maximum acceptable delay between the time of data generation and the time the information is dispatched at the receiver. Finally, with TRANSPORT_PRIORITY, a best-effort service is offered for information dispatch. However, a priority can be specified for queue management.

In order to deal with constrained devices, DDS also includes QoS mechanisms that limit the *resource usage*. The TIME_BASED_FILTER parameter allows an application to specify the *minimum* inter-arrival time between data, thus limiting the maximum generation rate. The middleware, in this way, does not flood them with bursts of data that they are not able to process (Fig. 3).

4.3 CoAP

The Constrained Application Protocol (CoAP) [22], recently standardized by the IETF, is a REST protocol specifically designed for constrained devices operating in low-power and lossy networks. CoAP takes strong inspiration from HTTP;

however, it has a reduced overhead and is optimized specifically for M2M communications, e.g., it uses the user datagram protocol (UDP). Following a client-server RESTful architecture, CoAP implements a request/response model with REST methods (GET/PUT/POST/DELETE). In a CoAP environment, constrained devices are usually servers that expose their sensors/actuators as dedicated resources, each one identified by a unique URI. On the other hand, clients interact with servers through the REST interface to retrieve the value of a sensor (GET) or to perform an action on an actuator (POST).

Constrained devices are usually battery powered, and they may be in a sleeping state for a certain amount of time. In order to handle sleepy nodes and optimize battery, CoAP allows the use of proxies. Moreover, the use of proxies is encouraged to implement caching or QoS support that cannot be directly implemented on constrained devices. Specifically, two different types of proxies are defined: forward and reverse. With the former one, clients are aware of the presence of the proxy and issue requests to the proxy with dedicated options that specify the required end point. The reverse proxy, instead, hides real end points and exposes remote resources as they were owned directly by the proxy. Clients issue normal requests to the reverse proxy, which is in charge of forwarding the requests/responses to/from the actual end point.

Finally, CoAP has been enhanced by introducing novel features that are specifically designed for IoT applications. Among the others, the observing resource feature is especially of interest [26]. This extension is based on the well-known observer design pattern, where "observers" (clients) register at a specific, known provider (server), for a specific "subject" (resource), communicating that they are interested in being notified whenever the subject undergoes a change in state. In this way, a client can register its interest for a resource hosted by a server; the server will then send asynchronous notification messages to all registered clients whenever the resource state changes. This approach minimizes the number of messages and ensures delivery of fresh updates.

Observing can be effectively combined with the use of proxies. In particular, if two clients are interested in the same resource, the proxy may act as an observer on behalf of the clients. In this case, the proxy registers on the remote end point and when it receives a notification, it forwards the update to all interested clients. This relieves the server from sending the same notifications multiple times and thus ensures scalability.

4.4 *CoAP QoS*

The CoAP standard does not include any feature that allows clients to specify the desired QoS in interacting with servers or in establishing an observe relationship. Considering the importance of QoS guarantees for use cases that require timely delivery of information, many research efforts have recently focused in introducing QoS mechanisms in CoAP. Among them, the majority focused on adding QoS support in the observing feature.

In [27], the authors propose to differentiate notifications by introducing delivery priorities. Specifically, a delivery option is introduced in each notification to prioritize the delivery of high priority messages through the network. However, no assurance is provided to CoAP clients, which are not allowed to set a notification period. In [28], a proxy-based architecture is proposed to overcome the limited capabilities of sensors. A virtualized proxy environment is introduced as a framework to offer differentiated services to groups of clients. Each group communicates with sensors through a different (virtual) proxy, which can implement custom functionalities in a transparent manner, e.g., a QoS policy to prioritize CoAP requests by group of clients.

When a resource is of interest for multiple clients, it may happen that each client has its own different QoS requirements. In fact, in the case of observing, one client might be interested in not receiving too many notifications within a short time period, if the state of a resource changes too often. On the other hand, another client might be interested in receiving notifications anyhow, in order to make sure that the end point is still working. To accommodate such requirements, the "Reusable Interface Definitions for Constrained RESTful Environments" Internet Draft [29] introduces, among others, the capability by a client to control the observe behavior by specifying a minimum period (p_{min}) and a maximum period (p_{max}) between two successive notification messages. Specifically, p_{min} and p_{max} set the maximum and minimum frequency, respectively, with which the client is willing to receive notification messages from the server, regardless of state changes. Managing multiple observe relationships for the same resource with heterogeneous QoS parameters is non-trivial and often cannot be performed directly by constrained devices that usually expose a simple interface to control observe, i.e., specifying only a fixed notification period, which must be the same for all possible clients [30]. To overcome such issue, the authors in [31] propose to deploy a proxy between clients and servers, which establishes an observe relationship with the server on behalf of all clients. Every time a periodic notification is received, and it is then forwarded to observers according to their respective QoS requirements, specified in terms of a minimum and maximum notification period. The selection of a unique notification period for the observe relationship with the actual server is performed through an algorithm that computes the period that fulfills all different p_{min} and p_{max} periods specified by the clients. The algorithm is designed to reduce both the load in the network and the energy consumed by devices.

4.5 MQTT

MQTT is a lightweight publish/subscribe messaging transport protocol [24]. By default, it runs over TCP/IP; however, it can be deployed on top of any protocol that provides ordered, lossless, bidirectional connections. The MQTT protocol is based on a central entity, called *broker*, which manages all the topics available in the network. In particular, when a client wants to distribute a message, it connects to the broker to publish the message under a certain topic. On the other hand, multiple clients connect to the broker and subscribe to topics. In this way, whenever a new message is published under a certain topic, the broker dispatches the same message to all the subscribers of such topic, achieving a one-to-many messaging distribution protocol. However, it is worth to mention that the broker treats every receiver independently.

The topic can be seen as a description of the information. However, MQTT is agnostic to the content of the messages and there is not a common semantic. How to use information is up to the clients; consequentially, there is the need of a previous knowledge between clients to achieve M2M communications.

4.6 MQTT QoS

MQTT provides QoS support in terms of different policies available to deliver a message from a sender to a receiver. In particular, MQTT allows three different QoS levels: *At most once delivery* (QoS 0), *At least once delivery* (QoS 1), and *Exactly one delivery* (QoS 2).

When a client publishes a message, it also defines the QoS treatment. When a message is published with QoS 0, it is delivered according to the capabilities of the underlying network. The receiver sends no response, and no retry is performed by the sender. The message arrives at the receiver either once, or not at all—see Fig. 4. With QoS 1, instead, the message arrives at the receiver at least once. A QoS 1 PUBLISH Packet has a packet identifier in its variable header and is acknowledged by a PUBACK Packet,—see Fig. 5. In particular, the sender selects an unused Packet ID and creates the message. Such message is stored on the sender side until the acknowledgment message for the same Packet Id is received. Storing of messages is exploited in order to manage particular circumstances in which TCP sessions fails, allowing a sender to retransmit unacknowledged packets whenever the client reconnects.

Finally, the QoS 2 is the highest quality of service, for use when neither loss nor duplication of messages are acceptable. There is an increased overhead associated with this quality of service because a two-step acknowledgment process is employed. This means that the first two messages are the same presented for QoS 1,

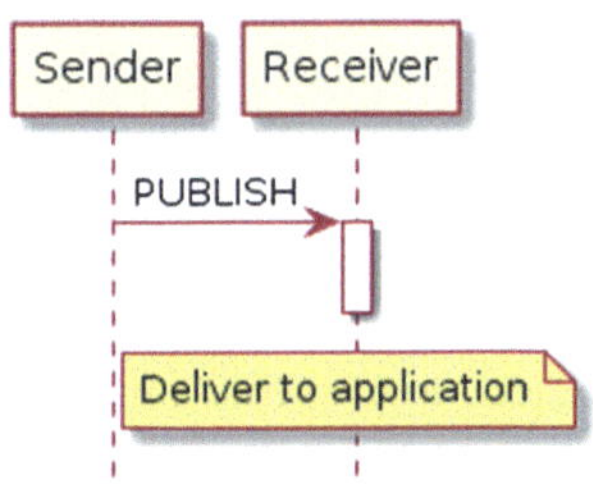

Fig. 4 MQTT—QoS 0

Fig. 5 MQTT—QoS 1

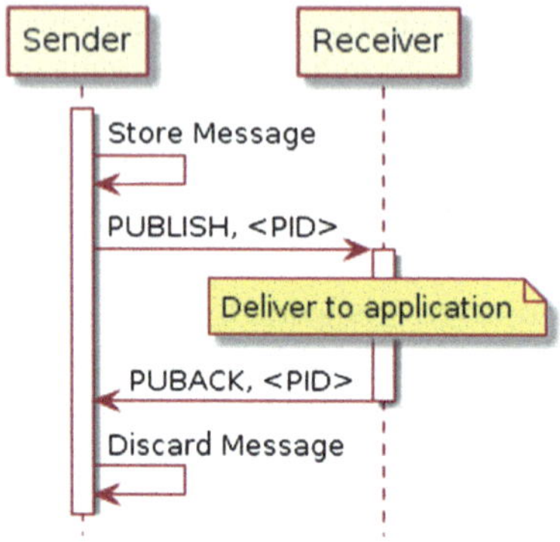

but they are followed by two other messages. This second exchange is exploited to acknowledge the receiver regarding the fact that the sender will not retransmit the origin message in the future.

4.7 oneM2M

OneM2M [32] is an international standardization effort that focuses on defining a standard architecture for IoT systems to support a wide range of applications and services on heterogeneous use cases. The oneM2M technical specifications define the architecture of a horizontal platform that can be deployed on various hardware and software to connect the myriad of devices in the field with applications. The overall oneM2M architecture is depicted in Fig. 6, where there are four different logical components, called nodes: application dedicated node (ADN), application service node (ASN), middle node (MN), and infrastructure node (IN). ADNs and ASNs are the IoT devices deployed in the field domain. ADNs are constrained devices that can perform simple actions, e.g., collect data or trigger actions on other sensors. ASNs, instead, are powerful devices that can implement more complex functionalities that are offered to other sensors or applications. To support the field domain, the oneM2M standard defines MNs, which are the M2M gateways deployed in the edge network, to provide services to applications deployed on ASNs and ADNs. Finally, in each oneM2M system, an IN is deployed in the infrastructure domain. An IN is similar to a MN except for the fact that it can also interact with other INs owned by external M2M systems.

From a functional point of view, oneM2M distinguishes between three different functional entities: application entity (AE), common service entity (CSE), and network service entity (NSE). An AE is an entity that implements an M2M application logic, and it is always connected to a CSE. A CSE, in fact, is a functional component that provides enhanced services to applications. Among the others, it is worth to mention the *data management* service, which is exploited to allow the exchange of data between AEs, and the *discovery* service, which enables

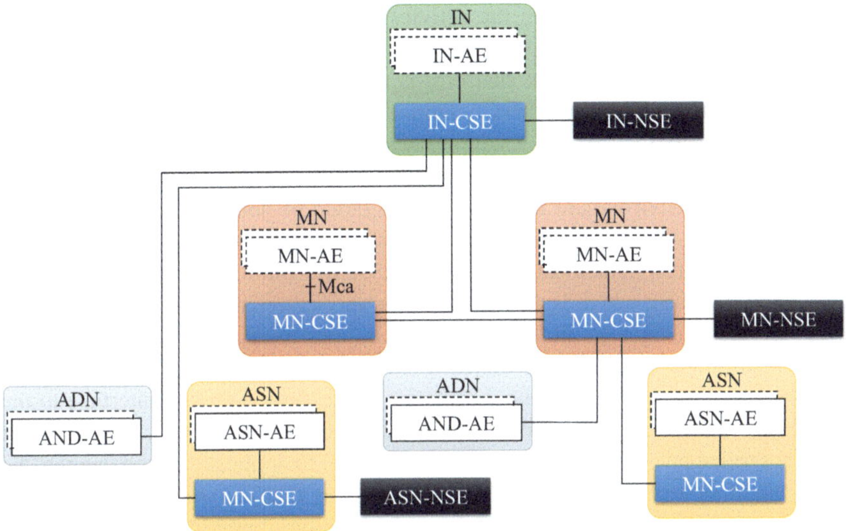

Fig. 6 oneM2M architecture

an AE to discover other AEs potentially connected to other remote CSEs. Finally, an NSE exposes services implemented into the underlying network and can be exploited to manage network-specific functions.

The data model of oneM2M is strictly resource-based; all services are exposed as resources, each one uniquely addressable through a URI. Such approach exposes a unified interface to applications, hiding the network and devices technical details. Moreover, it allows to interact with IoT devices through a standard REST interface that exposes a CRUD interface enhanced with the NOTIFY method. The latter allows CSEs to send asynchronous messages to other AEs regarding changes in a resource.

The oneM2M specifications include only a minimal support for QoS. Specifically, it allows only the possibility to specify a maximum response time for application requests, which is used merely to drop outdated messages. However, the NSE can be exploited to interact with the underlying layer to configure connections with QoS requirements, to manage devices or to trigger network operations, e.g., to wake up the device, or to establish a communication from the field domain toward the infrastructure domain. Moreover, IoT devices may communicate periodically; consequently, the NSE can be exploited also to optimize the operations in the underlying network, e.g., put devices to sleep for longer periods.

4.8 BETaaS

BETaaS [33] is an IoT platform for the development of M2M applications that is based on a distributed architecture rather than a centralized cloud-based infrastructure. Developed during the activity of the project *BETaaS: Building the Environment for the Things-as-a-Service*, a European project founded under the 7th Framework Programme, and the platform is available as open source software.

The BETaaS platform provides a unified framework for the development of M2M applications. It is designed with a layered structure that supports integration of existing systems and platform expandability. The platform relies upon a distributed runtime environment made of a *local cloud* of nodes that allows applications to access smart objects connected to the platform regardless of their technology and physical location. The core of the platform is the Thing-as-a-Service layer (TaaS), which is a content-centric service-oriented interface, deployed in a distributed manner on all the nodes, which hides to applications all the details of the distributed environment. Existing M2M systems can be integrated into BETaaS by implementing specific Adaptaters that expose a unified interface to the TaaS layer, translating the interface offered by the specific system. On the other side, the service layer provides a simplified interface to applications and enables the development of *extended services*, custom services that can run natively on the platform to extend the platform capabilities. In addition, it provides built-in support for non-functional requirements such as QoS, big data management, and security.

The BETaaS platform provides QoS support for applications. Considering the broad variety of applications that can be supported by IoT platforms, the implementation of QoS support is a non-trivial challenge. A classic approach adopted in different contexts starts from the definition of a standard QoS model to categorize QoS requirements into a predefined set of service classes [34]. In BETaaS, a simple schema composed by three service classes has been adopted: *real-time service* (applications with hard response time requirements), *assured service* (applications with soft response time requirements), and *best-effort service* (applications that do not require any assurance). At the same time, flexibility is guaranteed allowing applications to customize their requirements through a dynamic negotiation procedure within the selected service class. The negotiation is performed at the time of deployment following a two-stage procedure: first, the application specifies the QoS parameters required for each service, then the platform internally negotiates the thing services, the basic services offered by things, required to fulfill application requirements. The QoS negotiation protocol allows an application to specify the required service level.

In order to enforce and monitor the negotiated QoS requirements, a QoS framework has been defined and implemented in the platform in order to ensure an efficient management of resources while guaranteeing the fulfillment of the agreements with applications. The framework is based on a two-phase procedure, namely *reservation* and *allocation*—see Fig. 7. The reservation phase is handled by a *broker* at the time of application installation. The broker manages the QoS negotiation, performs admission control, and, most importantly, manages resource reservation. To this aim, a centralized component is introduced that has the knowledge of all the connected devices and of all applications' requests. The allocation phase, instead, is managed by local *dispatchers* implemented on each gateway. The dispatcher performs allocation of resources at time of invocation

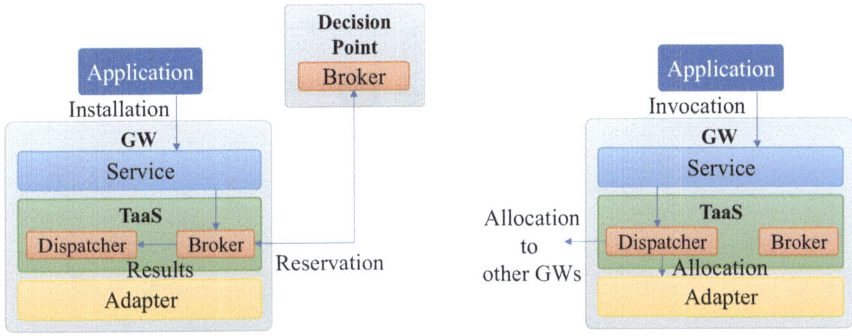

Fig. 7 BETaaS reservation and allocation

based on the results of the reservation phase. Resource allocation can manage the resources following different optimization goals. An example algorithm that optimizes the energy efficiency of battery-powered smart sensors is described in [35]. Regardless of the allocation, the platform does not implement any mechanism to configure QoS guarantees at the network layer according to the allocation. For an exhaustive description of the QoS framework introduced in the platform, the interested reader is referred to [36].

4.9 IoT@Work

IoT@Work is an IoT platform specifically designed for applications and processes in the manufacturing domain. The platform specifically focuses on providing reliable communication and security guarantees. To this aim, the platform includes by design functionalities to configure the underlying networks for enforcing the QoS guarantees required by applications.

The platform adopts a publish/subscribe message exchange schema to allow exchange of soft real-time streams between devices and applications. Core of the platform is the communication plane, which is responsible for enforcing QoS. Specifically, it manages network resources by creating per application "virtual networks" called slices. Each slice is created on demand to satisfy the QoS requirements of the application to which it is assigned. Slices fulfill QoS by relying on a traffic labeling mechanism similar to MPLS or VLAN. This cross-layer approach interacts with the network stack in order to configure paths and QoS guarantees (e.g., priorities and bandwidth) on the network devices.

The IoT@Work implements such interactions through the slice manager system, as shown in Fig. 8. The slice manager is responsible for providing a network abstraction to applications that are aware only of the communication end points. The slice manager must also enforce the QoS properties into the network by configuring paths between applications and end points. Applications interact with

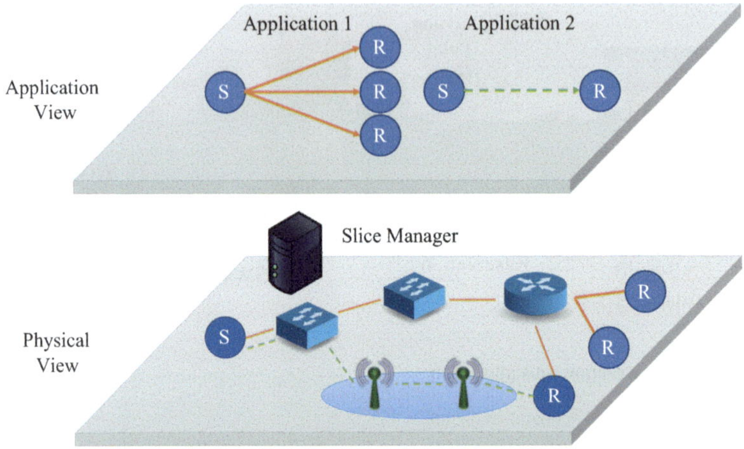

Fig. 8 IoT@work slice manager

the slice manger to specify the QoS requirements through SLA contracts. The slice manager, however, is a centralized entity, and it does not perform any resource allocation operation.

5 Future Research Directions

Although network-level QoS support for IoT networks is being made available not only as protocol specification but also in several real implementations, such features are mainly unexploited by current IoT platforms, which still do not include a complete cross-layer QoS support.

Among all the IoT platforms, the few that support QoS by design do not include all-embracing QoS features, and instead focus only on resource management or network management. In the case of the BETaaS platform, for instance, the QoS capabilities offered to applications do not leverage QoS features provided at the network layer. QoS is enforced through proper management of resources and appropriate allocation of requests from applications. However, the platform does not consider the contribution of the network in the overall service level experienced by applications. However, the contribution of the network to the overall service level may not be negligible, especially in network deployments such as large multi-hop wireless networks that potentially are characterized by noticeable communication latency and loss. The IoT@Work platform, on the other hand, focuses mainly on managing network resources, without considering proper management of application requests. This might result in an unbalanced use of IoT devices leading to congestion or violation of QoS requirements.

One significant future research direction will be the design of solutions to offer IoT applications a reliable QoS support through a cross-layer approach that enforces QoS requirements both at the application level and the network level. Specifically, the following scientific and technological challenges are envisioned:

- Definition of a *cross-layer reference architecture for QoS support* in IoT platforms in order to exploit the QoS functionalities offered by communication networks. Solutions to guarantee proper integration of different network protocols will be required in order to handle different QoS models and interfaces currently under standardization in different communication architectures.
- Development of *QoS enforcement algorithms for cross-layer resource management*. Specifically, the definition of novel solutions (or modifications to existing algorithms) in order to manage resources from both the service and network layers will be necessary to enforce fine-grained requirements from applications. One example is the definition of algorithms to translate high-level application requirements into low-level requirements for service and network layers, respectively.

References

1. Z. Chen, C. Wang, Use Cases and Requirements for using Track in 6TiSCH Networks, IETF Internet Draft
2. K. Pister, P. Thubert (eds.), S. Dwars, T. Phinney, Industrial Routing Requirements in Low-Power and Lossy Networks, RFC 5673
3. Wenye Wang, Xu Yi, Mohit Khanna, A survey on the communication architectures in smart grid. Comput. Netw. **55**(15), 3604–3629 (2011)
4. D. Niyato, L. Xiao, P. Wang, Machine-to-machine communications for home energy management system in smart grid. IEEE Commun. Mag. **49**(4), 53–59 (2011)
5. IEEE Standard Communication Delivery Time Performance Requirements for Electric Power Substation Automation, in *IEEE Std 1646-2004*
6. V.C. Gungor et al., Smart grid technologies: communication technologies and standards. IEEE Trans. Industr. Inf. **7**(4), 529–539 (2011)
7. Ó. Gama et al., Quality of Service Support in Wireless Sensor Networks For Emergency Healthcare Services, in *2008 30th Annual International Conference of the IEEE Engineering in Medicine and Biology Society* (IEEE, 2008)
8. N. Chevrollier, N. Golmie, On the Use of Wireless Network Technologies in Healthcare Environments. White Paper—U.S Department of Commerce, July 2005
9. M.M. Baig, H. Gholamhosseini, Smart health monitoring systems: an overview of design and modeling. J. Med. Syst. **37**(2), 9898 (2013)
10. D. Chen, M. Nixon, A. Mok, *WirelessHART: Real-Time Mesh Network for Industrial Automation* (Springer Publishing Company, Incorporated, 2010)
11. M. Collotta, G. Pau, G. Scatà. Deadline-aware scheduling perspectives in industrial wireless networks: a comparison between IEEE 802.15.4 and Bluetooth. Int. J. Distrib. Sens. Netw. 2013 (2013)

12. P. Soldati, H. Zhang, M. Johansson, Deadline-Constrained Transmission Scheduling and Data Evacuation in WirelessHART Networks, in *2009 European Control Conference (ECC)*, Budapest (2009), pp. 4320–4325
13. P. Thubert, An Architecture for IPv6 over the TSCH mode of IEEE 802.15.4, IETF Internet Draft
14. Q. Wang, X. Vilajosana, 6top Protocol (6P), IETF Internet Draft
15. S. Shah, P. Thubert, Differentiated Service Class Recommendations for LLN Traffic, IETF Internet Draft
16. M.R. Palattella et al., On-the-fly bandwidth reservation for 6TiSCH wireless industrial networks. IEEE Sens. J. **16**(2), 550–560 (2016)
17. N. Accettura et al., Decentralized traffic aware scheduling in 6TiSCH networks: design and experimental evaluation. IEEE Internet Things J. **2**(6), 455–470 (2015)
18. J. Decuir, Introducing Bluetooth smart: part II: applications and updates. IEEE Consum. Electron. Mag. **3**(2), 25–29 (2014)
19. J. Nieminen et al., IPv6 over Bluetooth Low Energy, RFC 7668, October 2015
20. J. Nieminen et al., Networking solutions for connecting bluetooth low energy enabled machines to the internet of things. IEEE Netw. **28**(6), 83–90 (2014)
21. C.F. Hsu, C.Y. Liu, An adaptive traffic-aware polling and scheduling algorithm for Bluetooth Piconets. IEEE Trans. Veh. Technol. **59**(3), 1402–1414 (2010)
22. Z. Shelby, K. Hartke, C. Bormann, The Constrained Application Protocol (CoAP), RFC 7252, June 2014
23. Data Distribution Service (DDS), Version 1.4 (2015)
24. MQTT Version 3.1.1. Edited by Andrew Banks and Rahul Gupta. 10 April 2014. OASIS Committee Specification Draft 02/Public Review Draft 02
25. The Real-time Publish-Subscribe Protocol (RTPS) DDS Interoperability Wire Protocol Specification, Version 2.2 (2014)
26. K. Hartke, Observing Resources in the Constrained Application Protocol (CoAP), RFC 7641, September 2015
27. A. Ludovici, E. Garcia, X. Gimeno and A. Calveras Augé, Adding QoS Support for Timeliness to the Observe Extension of CoAP, in *2012 IEEE 8th International Conference on Wireless and Mobile Computing, Networking and Communications (WiMob)*, Barcelona (2012), pp. 195–202
28. E. Mingozzi, G. Tanganelli, C. Vallati, CoAP Proxy Virtualization for the Web of Things, in *IEEE International Conference on Cloud Computing Technologies and Science (CloudCom)*, Singapore, 15–18 December 2014
29. Z. Shelby, M. Vial, M. Koster, Reusable Interface Definitions for Constrained RESTful Environments, draft-ietf-core-interfaces-04, October 2015
30. M. Kovatsch, O. Bergmann, C. Bormann, CoAP Implementation Guidance, draft-ietf-lwig-coap-03, January 2016
31. G. Tanganelli, E. Mingozzi, C. Vallati, M. Kovatsch, Efficient Proxying of CoAP Observe with Quality of Service Support, in *Proceedings of the IEEE 3rd World Forum on Internet of Things (IEEE WF-IoT 2016)*, Reston (VA), USA, December 12–14, 2016
32. TS-0001-oneM2M-Functional-Architecture, -V2.10.0 (2016)
33. C. Vallati, E. Mingozzi, G. Tanganelli, N. Buonaccorsi, N. Valdambrini, N. Zonidis, B. MartÃ-nez, A. Mamelli, D. Sommacampagna, B. Anggorojati, S. Kyriazakos, N. Prasad, F. Nieto De-Santos, O. Barreto Rodriguez, BETaaS: A Platform for Development and Execution of Machine-to-Machine Applications in the Internet of Things, in *Wireless Personal Communications*, Published on line 13 May 2015
34. R. Liu et al., M2M-Oriented QoS Categorization in Cellular Network, in *Proceedings of the 2011 7th International Conference on Wireless Communications, Networking and Mobile Computing* (2011)

35. G. Tanganelli, C. Vallati, E. Mingozzi, Energy-Efficient QoS-aware Service Allocation for the Cloud of Things, in *Proceedings of the IEEE Workshop on Emerging Issues in Cloud (EIC 2014)—co-located with IEEE CloudCom 2014*, Singapore, 15–18 December 2014
36. E. Mingozzi, G. Tanganelli, C. Vallati, A Framework for QoS Negotiation in Things-as-a-Service oriented Wireless Communications, in *Proceedings of the 4th International Conference on Wireless Communications, Vehicular Technology, Information Theory and Aerospace & Electronic Systems (Wireless VITAE 2014)*, Aalborg, Danemark, 11–14 May 2014

Wireless Sensor Networks-IoT Infrastructure

Azadeh Zamanifar

Abstract In this chapter, we consider IP-based mobile wireless sensor networks as one of the most important infrastructure for IoT. We first explain the feature and challenges for this context. In IP-based mobile wireless sensor networks, maintaining the connectivity of mobile node to the network is the real challenge. This is an important issue mostly for time-critical applications like health care. The other important factor is reducing hand-off cost during inter-PAN and intra-PAN movement. This results in reducing power consumption of mobile nodes which is the other important issue in this context.

Keywords Wireless sensor networks · Mobile nodes · PAN movements · IP-based node

1 IoT Applications and Wireless Sensor Networks' Rule

The number of advanced digital devices has been dramatically increased in recent years. The Internet of Things (IoT) is known as the third revolution in information technology [10]. Because of the emergence of IoT, IPv6 low-power personal area networks (6LoWPANs) have recently found renewed interest. However, mobility support in 6LoWPANs is still in its infancy for large-scale IP-based sensor technology in future IoT. The Internet of Things paradigm has recently attracted a vast amount of attention in both academia and practice [2, 7, 20, 27]. Wireless sensor networks (WSNs), as a fundamental enabling technology of IoT, integrate a number of spatially distributed autonomous sensors into a network which cooperatively pass their data through wireless communication, in the same or different networks [12, 14, 19, 22].

Health care as an important 6LoWPAN IoT application keeps continuous monitoring of patients' vital signs while the patients are moving [22]. Due to the importance of health-care services, sophisticated mobility management mechanisms are

A. Zamanifar (✉)
Shahid Beheshti University, Daneshjoo Blvd., Shahid Shahriyari Sq.,
Yaman St., Chamran Hwy., Tehran, Iran
e-mail: a_zamanifar@sbu.ac.ir

R.R. Yager and J. Pascual Espada (eds.), *New Advances in the Internet of Things*,
Studies in Computational Intelligence 715, DOI 10.1007/978-3-319-58190-3_10

required to maintain connectivity between patients' nodes and the hospital/nursing house network to monitor their exact locations. These mechanisms should also support fault tolerance while optimizing energy consumption of the sensing devices [15, 21].

In a typical health-care system, various mobile sensor nodes are employed for monitoring different vital parameters of a patient. Some of the most commonly used sensors in unobtrusive ubiquitous health monitoring are as follows: pulse oxygen saturation sensor, blood pressure (BP) sensor, serum glucose level, electrocardiogram (ECG) sensor for monitoring heart activity, electromyogram (EMG) sensor for monitoring muscle activity, electroencephalogram (EEG) sensor for monitoring brain electrical activity, temperature sensor, core body and skin temperature, and breathing sensor for monitoring respiration [17]. IoT health care provides two-directional communications between a mobile node(s) and the remote server which is not possible in traditional health-care systems.

IoT health-care system has the challenges of typical wireless sensor networks (WSN), including power constraints. The limited power of sensor nodes is one of the main concerns in every wireless sensor networks. Also, safety critical applications, such as health-care system, have some significant challenges which are not the primary issues of conventional sensor networks. The main concern in a pervasive health-care system is providing a real-time response by which emergency situations, e.g., heart attacks or sudden falls, can be identified and reported in a short period [1]. Another important issue in health-care systems is continuous connectivity of the mobile node(s) to the network.

To cope with some challenges of the health-care system, including limited power consumption, static nodes are deployed in the monitoring area. Static nodes are less power constrained compared to the mobile sensor nodes. Without static nodes, mobile nodes should directly communicate to the access point/gateway. Thus, a considerable amount of mobile nodes' power may be consumed due to the distance between the mobile nodes and the access point(s) [4].

There are two general methods for mobility management of mobile IP-based sensors: (1) deploying static nodes to handle the movement of mobile nodes, that is applicable in poor infrastructure environments and (2) direct communication of mobile node(s) to access point(s) that is applied in rich infrastructures. In the latter case, multiple access points are deployed to reduce the energy consumption of mobile nodes by decreasing the distance between access points and mobile nodes [4]. However, rich infrastructure is not always available since it is not cost effective.

2 Mobility Management

In typical applications of IoT, it is common that things frequently move. This necessitates the existence of an appropriate mobility management protocol in order to keep things connected to the Internet via a gateway [29]. Mobile health care has attracted a lot of attention as it deals with improving the quality of human health, well-being and ease of use for people [5, 14, 15]. It has been predicted that in the subsequent decades, the health-care environment, which is currently hospital-centered, will be transformed to hospital–home-balanced in 2020th, and eventually to home-centered in 2030th [16]. Providing efficient mobility management has the foremost role in designing a sophisticated system such that there's always been a connection between the mobile node(s) and gateway [6, 18].

In WSNs, there are two types of mobility: macro-mobility and micro-mobility. The movement of nodes between different network domains is known as the macro-mobility while in micro-mobility, the nodes move within the same network domain. Several mobility management solutions, based on a single-hop or multi-hop mesh routing, are briefly described in this section. Furthermore, the mobility management protocols are either network based or node based. It is defined based on the fact that which entity takes the mobility overhead. As it is mentioned above, the routing has a direct impact on mobility management. Mesh-under routing protocols do not need any data buffering in intermediate nodes. Thus, it consumes less power and delay. But if the network is unreliable and only a packet is lost, all the fragmented packet of the messages must be retransmitted from source to destination. In route-over solutions like RPL, the routing is done in network layer which makes it more reliable as each intermediate node receives all frames and reassembled it in the network layer.

2.1 *Movement Direction Prediction Approaches in Overall*

The existing movement prediction approaches in IP-based sensor networks with mobile sensor nodes are typically based on hardware-specific methods. Two primary ways in this context use angle of arrival (AoA) [11, 23–26]. However, these methods require specific hardware facilities which are not always available in real applications. While the hardware devices, such as the directional antenna, are prone to failure, the corresponding methods do not consider self-healing mechanisms in case of static node(s) failure. Thus, the failure of a static node results in network disconnectivity of the mobile node. Furthermore, if the direction of a mobile node suddenly changes, the methods fail to determine the correct movement direction. Besides, estimating the angle and distance between the mobile and static nodes is highly sensitive to noise as it relies on RSSI. In link quality indicator (LQI)-based methods, the direction of movement can be detected only by broadcasting a message and measuring LQI of the messages which are exchanged between the mobile and neighboring static nodes. This method suffers from a huge amount of message exchanges which decrease the lifetime of mobile nodes while causing an inappropriate amount of delay [3, 9, 11, 13, 30].

There is a machine-learning-based approach for predicting the movement direction of mobile nodes in IP-based sensor networks [28]. This is the first work which uses machine learning for movement prediction, and consequently on reducing hand-off cost, in these networks. Predicting movement direction, based on a machine learning model, eliminates the need for online movement direction detection methods which imposes computation and communication overhead during handoff. The new approach, so-called Distribute Self-Healing Movement Prediction in IOT (abbreviated as DSHMP-IOT), uses hidden semi-Markov model (HSMM) to predict the movement direction of the mobile node(s). This is also the first approach which considers the failure of static node(s) in mobile IP-based wireless sensor networks.

2.2 *DSHMP: Learning-Based Movement Direction Prediction Approaches*

In a monitoring area, such as an elderly house, we build IP-based mobile WSN with n mobile nodes and m static sensors. The area is divided into $\log m$ equal-sized cells. The proposed tree structure, DSHMP-Tree, is built such that each static node at the center of each cell acts as a particular leaf of the tree. The static nodes which serve as predecessors of the DSHMP-Tree leaves are located in predefined locations between cells to connect the leaves to the gateway. The static nodes are deployed as a binary tree to reduce handoff during movements of the mobile object(s). Every mobile node sends its data through the nearest static node (which is called candidate node) to the gateway. The candidate node periodically checks the mobile node's status to be informed about its movement. Figure 1 reveals steps for the proposed movement prediction approach. As shown in the figure, during the training stage, the patient's tracking data are collected and forwarded to the gateway via the DSHMP-Tree where a hidden semi-Markov model (HSMM), presented as two tables, is constructed. To distribute the model over the static nodes, the tables are further divided into sub-tables, each of them carrying data required for predicting the direction of the patient movement regarding each cell of the area. These sub-tables are forwarded through DSHMP-Tree and distributed over the static leaf nodes, positioned at the center of each cell.

The approach has four main parts including initial setup (which consists of tree construction and collecting movement data), movement prediction, mobility management and self-healing that are described in the following sections.

During the training or data collection phase, there is no movement direction prediction technique, and the direction of the mobile node's movement is detected via RSSI and LOI. Each sub-trees including eight leaves is considered as a single hidden state of HSMM. Each cell corresponds to one output or observation. We analyze different sub-tree sizes. In our case, the sub-tree with eight leaves gives us the better movement direction accuracy results. The reason to choose a sub-tree as a hidden state is that in a nursing house, each part of the building is related to the activities that patient do in the specific time of the day. Thus, this logical division shows the similar behavior of the person during staying in a state. For each leaf of the tree, an individual model is sent after the training phase from the gateway and forwarded as an appropriate table to the node. During the test phase, whenever the patient is about to move from its current cell to another cell, the movement prediction function is

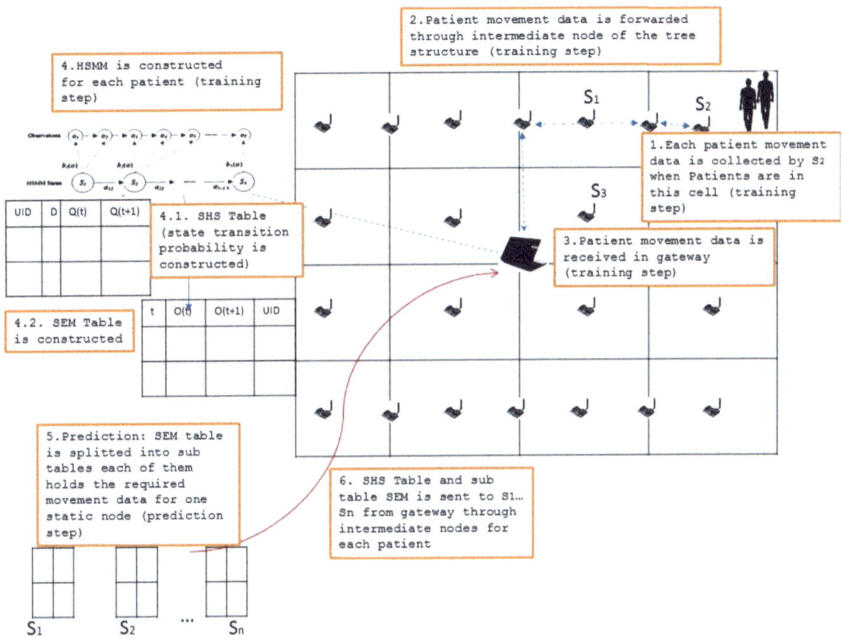

Fig. 1 The proposed DSHMP-IOT movement prediction stages based on DSHMP-Tree network scheme

invoked by looking up in the corresponding table of the current candidate node, and the movement direction is identified.

Initial setup of the DSHMP-Tree is described in Sect. 2.3 should be firstly built. Subsequently, in Sect. 2.4, the movement prediction model should be generated and distributed according to the constructed tree. In Sect. 2.5, mobility management is described. Finally in Sect. 2.6, our self-healing mechanism in the case of failure of the static node(s) is explained .

2.3 *Initial Network Setup*

At the initial network setup, firstly we put static nodes DSHMP-Tree, detailed in Sect. 2.3.1. Then, build the tree in Sect. 2.3.2. Subsequently, collecting movement data is described in Sect. 2.3.3.

2.3.1 Static Node Placement

The monitoring area is assumed to be a Cartesian plane with coordinates X and Y where the origin is left down corner of the area. Parameter k holds the current tree level, and L holds the cell's length, which could be in square meters. At first, the

Table 1 IPv6 address structure

96 bits	16 bits	8 bits	8 bits
Global routing prefix	Sub PANID	Tree	NodeID

leaves of the tree which are placed at the middle of each cell are located. To build the parent of each two child leaves, the x location of the parent node is the middle of its children x values, and it's y value is the y value of the children plus $L/2$. Subsequently, the static nodes at higher levels are positioned.

After the placement of nodes, each node finds its parent and its children. The address structure of static and mobile node is shown in Table 1. The *tree* part of the addresses in each level of the DSHMP-Tree is different. The root of the tree sends the address of its children. The *tree* part of the children address is twice the *tree* part of its parent address for its left child and twice the *tree* part of its parent address plus one for its right child. Then, each child that receive its address sends the address of its children in a similar manner. This is done in a recursive manner until the leaves of the tree set their IDs.

2.3.2 DSHMP-Tree Construction

The proposed scheme specifically helps us during handoff caused by patient's movement and also facilitates routing of mobile node data to the gateway. To build DSHMP-Tree, the monitoring area is divided into equal-sized cells. In other words, the nursing house is considered as a set of cells with length L.

While the leaves of the DSHMP-Tree are responsible for handling the movement of mobile nodes and receiving/forwarding data of mobile node(s) to their parents, the non-leaf nodes act as intermediate devices to receive and forward data between their parents and children. As an example, in Fig. 2, for simplicity, we use simple numbers and letters for specifying the nodes of DSHMP-Tree. With the same purpose, in remaining parts of the chapter, we use cell ID and corresponding leaf static node's ID, alternately. To manifest the forwarding process in DSHMP-Tree, consider Fig. 2 where the mobile node is in cell 1. Obviously, the static node 1 is the candidate node for the mobile node which forwards data d of the mobile node to node (1,2) which forwards it to node A. Subsequently, the data will be forwarded to node A' which will send it to node A''. Then it is relayed to node R_1 which will send it to the root of the tree. As shown in Fig. 2, we assume that the monitoring area is divided into n equal-sized cells.

In a monitoring area, e.g., an elderly house with M square meters area, we assume n cells each of them has l length. As each static sensor can see its neighbors, l equals to $1/(2 \times \sqrt{2})$ of the leaf node's communication range. This is shown in Eq. 1.

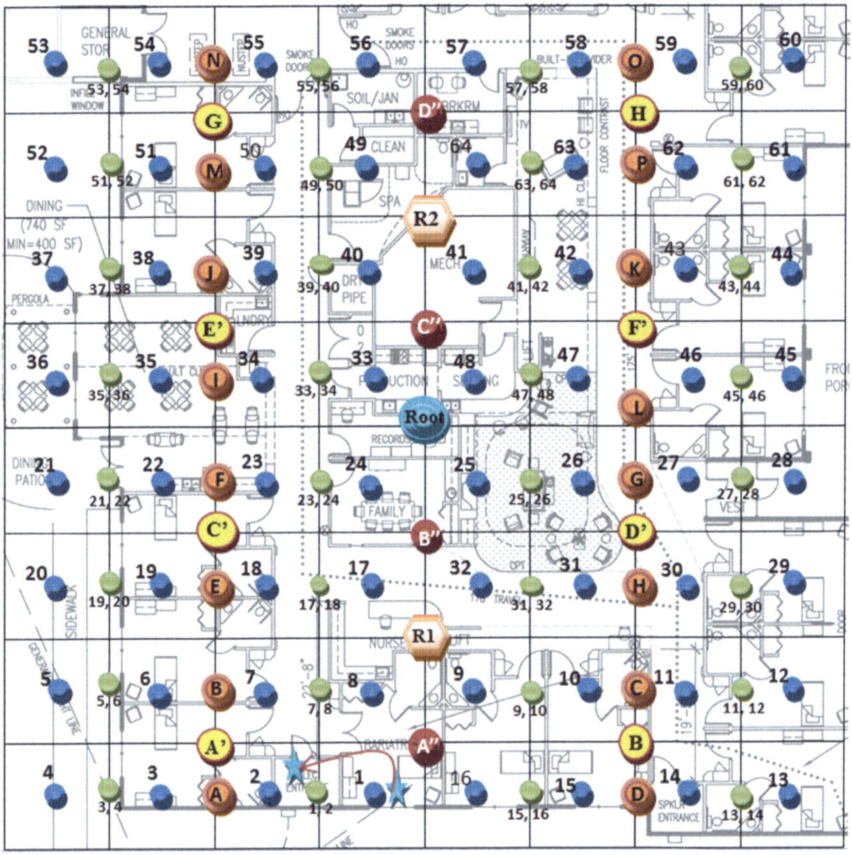

Fig. 2 Initial setup: tree construction

$$M = n \times l \times l \tag{1}$$

The DSHMP-Tree of Fig. 2 with 64 leaf nodes and 127 total static nodes is depicted in Fig. 3.

2.3.3 Collecting Movement Data

The data that are collected in the training step during a particular period have the following features: the time of the day, duration in which a patient stays in a cell and the current static sensor ID at the center of the cell in which the patient is located. During the training or data collection phase, there is no movement direction prediction technique, and the direction of the mobile node's movement is detected via RSSI and LOI.

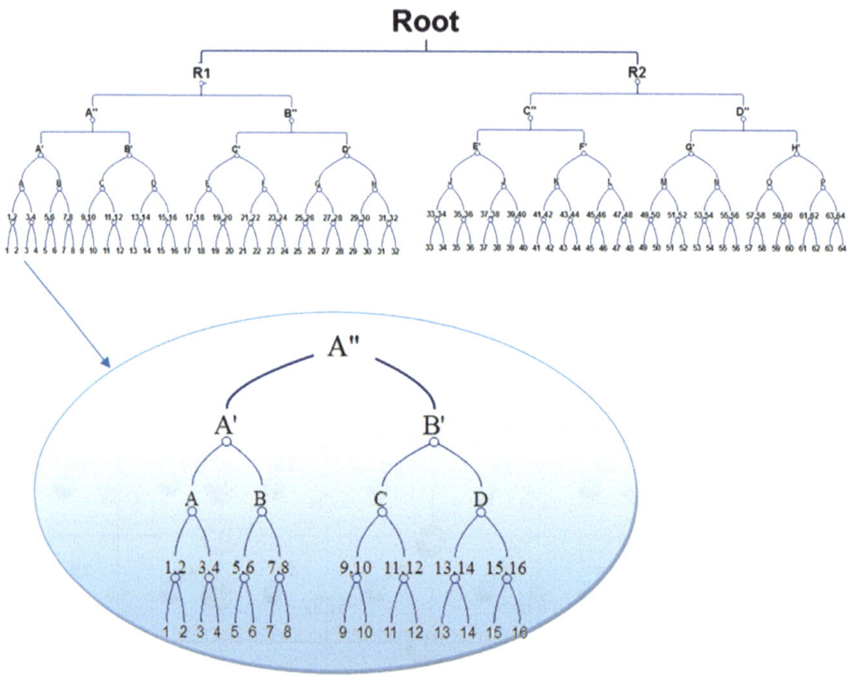

Fig. 3 DSHMP-Tree for the monitoring area of Fig. 2

2.4 Movement Prediction

In Sect. 2.4, we detail how movement prediction is made. In Sect. 2.4.1, it is described how the model is constructed. In Sect. 2.4.2, it is depicted how the generated data models are distributed after training phase over the leaves of the DSHMP-Tree. The movement prediction is achieved by modeling data that are collected from the movement of a given patient during the training phase. We have applied HSMM.

2.4.1 Constructing Movement Prediction Model

We train the model with syntactic movement data that we collect from simulating our network in Cooja environment [8]. The data contain sequences per patient's movement at different times of day in an elderly home. To achieve this, whenever patient moves from one cell to another one, the data of this movement are sent to the gateway via DSHMP-Tree. We also log the total duration of staying in the corresponding state. The current and next states that are the IDs of the great-grandparent of current and next sensors are also sent to the gateway. The minimum duration of staying in a cell can be obtained based on the logs of the movements. This minimum duration becomes the time step during model training. Furthermore, the total duration of staying in a state is also sent to the gateway by the root of each sub-tree (states).

As mentioned before, each mobile node sends its data via the nearest static node (which is called the candidate node) to the gateway. Thus, by the proposed model, when the mobile node is about to move, according to the trained model, the mobile node will go to the most probable cell in its neighborhood. Thus, it does not require communication between the candidate and other nodes in the vicinity, nor it requires a directional antenna.

2.4.2 Distributing Movement Model

In this section, the structure of the model, as well as the way for model distribution, is explained. *SHS* (semi-Markov hidden states) and *SEM* (semi-Markov emission) tables at the gateway have the following attributes:

1. *SHS Table*: (current hidden state, next hidden state, duration, user ID)
2. *SEM Table*: (user ID, next cell, time step, current cell): This table determines the next output (sensor ID that is centered in one of the neighboring cells).

After the model is constructed at the gateway, the corresponding *SEM* sub-table is forwarded to the relevant leaf node. For each leaf node *a*, its assigned sub-table contains those rows of the *SEM* table for which the *currentCell* column value is *a*.

2.5 *Mobility Management-Early Stages*

In this section, we only focus on the part of the mobility management that is related to movement prediction. Table 2 depicts the mobility management algorithm (only movement detection, identifying movement direction and recovery from false prediction parts). Each subsection describes one part of the above algorithm, related to parts of this pseudo code.

2.5.1 Movement Detection

To detect the movement of a mobile node, the candidate node sends the mobile node a message in the specific interval and computes RSSI. Whenever the RSSI of the message comes below a threshold, candidate node detects that mobile node is about to move (Line 1 of Table 2). It sends the elapsed time step to the next probable cell (new candidate node).

Table 2 Mobility management algorithm (movement detection, identifying movement direction, recovery from false prediction)

```
1. PrevCandidate node detects the movement of mobile node
2. PrevCandidate node sends NN message to predicted candidate node
3. Predicted candidate node sends CandidateReq message to mobile node
4. If Mobile node receives CandidateReq message then
5.     Mobile node sends CandidateRes message to predicted candidate node i
6.     i sends NNACK to previous candidate node
7.   Else
8.     It broadcasts FindCandid message containing (PevCandidatenodeID, PredictedCandidatenodeID)
9.     Any static node i that receive the FindCandidReq message from mobile node calculate RSSI
10.     If the calculated RSSI is above some threshold
11.         they send CandidateResponse Message to mobile node
12.         i:the static node with the highest RSSI sends NNACK to previous candidate node
13.         PrevCandidate node send the previous and current hidden state (room) to i
14.     EndIf
15. EndIf
```

2.5.2 Identifying Movement Direction

Once a candidate node detects the movement of the mobile node(s) (Line 1), it sends a *NN* message to the predicted candidate node (Line 2) to wake up it. The current candidate node also sends the *timestep* to the predicted candidate node. When the predicted candidate node receives the *NN* message, it sends a *CandidateReq* message to the mobile node (Line 3). If the mobile node receives the *CandidateReq* message, it sends *CandidRes* to the predicted candidate node which sends a *NNACK* message to the previous candidate node (Lines 5–6).

2.5.3 False Prediction Recovery

To compensate false prediction, DSHMP-IOT contains a recovery mechanism, depicted in Table 2. If the mobile node does not receive a *CandidateReq* message during a specific period, it broadcasts a *FindCandidReq* message containing the ID of its previous candidate node (Lines 7–8). A neighboring static node which receives this message, considering that the RSSI of the message is above the certain threshold, sends *FindCandidRes* to the mobile node. At the mobile node, the node ID with max RSSI is chosen (among the messages that are received) to be a new candidate node of the mobile node and sends a *NNACK* message to the previous candidate node (Lines 9–12). The previous candidate node sends the previous and current hidden state (room) to the newly identified candidate node as these data are required for predicting next cell in the next movement of the mobile node (Line 13).

2.6 *Self-healing*

Static nodes may fail or malfunction due to the battery problem. The communication link could also be broken. This results in disconnectivity of the mobile node from the gateway. Furthermore, since static leaf nodes in our proposed approach maintain part of the movement prediction model, failure of static node(s) may cause loss of corresponding data. To recover from static node failure, we proactively determine a substitute node for each static node, and we also maintain a duplicate of movement prediction model in each substitute node. In the following subsection, we firstly describe how substitute node is chosen proactively for each static node in the tree and then we will describe how to detect the failure of a static node and its restoration mechanism.

2.6.1 Determining Substitute Node

For each node of DSHMP-Tree, we proactively choose substitute node from its neighbors at initial start-up of the network or when the network configuration has changed. If a non-leaf node fails, its failure is detected, and the substitute node is responsible for restoring the network after the failure. To determine substitute nodes, the following algorithm is proposed:

1. If only one child y of a given node x is in the range of the parent of x, that child is chosen to be the substitute of its parent in case of failure.
2. If both children of a given node x are in the range of the parent of x, the grandchild which is nearer to its grandparent is chosen to be the substitute of x; if both children are located in equal distance to the grandparent, the node with smaller ID will be chosen. For example, the substitute for node A could be any of nodes (1,2) or (3,4) as both of them are in equal distance to A' which is the parent of node A. As another example, the substitute for node R'_1 is R_2 which is nearer to *root*.
3. The substitute for leaf nodes which are located at the center of each cell is their parent. For example, the substitute for node 1 is node 1,2.

To provide substitute node for each static node, static nodes must have different communication ranges. To this end, static nodes at the upper level need to have broader range comparing to the nodes at the lower level. The wider range of upper-level static nodes prevents the lower-level static nodes or mobile nodes disconnected from the gateway.

2.6.2 Detection and Recovery from Failure of Static Nodes

To handle the failure of a given static node, as shown in self-healing algorithm in Table 3, each node in the tree must be monitored by its parent (Line 1). Because of our automatic addressing scheme, these are known at initial setup. The node that detects the failure of its neighbor, and it is marked as substitute of the failed node, sends

Table 3 Algorithm for self-healing of static nodes

Self-healing algorithm
HEAL(node A)
{
1. Node *A* periodically send ALIVE message to parent(A)
2. If parent (A) do not receive ALIVE message from node A
3. node B= (substitute of A)
4. B broadcasts RECOVER message
5. child(parent(A))=B
6. child(B)=sibling(B)
7. parent (A) , B, child(B) Recalculate substitute node
8. parent (child (B))=B
9. parent(B) =parent(A)
10. If B is LEAF or child (B) is LEAF
11. Send MovementPredictionData(B) to substitute of B
12. Send MovementPredictionData(child(B)) to B
13. Endif
14. Endif
}

the *RECOVER* message including its ID to the parent of failed node and its sibling (Lines 3–4). These nodes update their tables as they receive this message (Lines 5–6). When the parent and child of the failed node receive *RECOVER* message, they also recalculate their new substitute node in the case that failed node is their current substitute node (Line 7). New substitute node also updates its parent/child tables accordingly (Lines 8–9). If the substitute node of the tree leaves is changed, the *EM* table of the tree leaves must be sent to their new substitute nodes (Lines 11–12). The algorithm of this part is shown in Table 3.

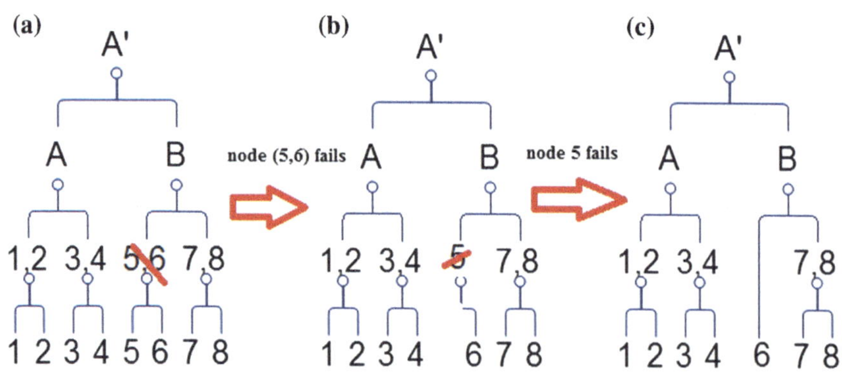

Fig. 4 **a** Node (5,6) fails, **b** Restoring network after node (5,6) fails, **c** Restoring network after node 6 fails

Figure 4 shows an example of network restoration after two static nodes (5,6) and 5 fail. According to the algorithm, node 5 becomes the father of node 6. If node 5 fails, node *B* becomes the father of node 6. To sum-up, the substitute for leaf nodes is their parent, and the substitute for the intermediate node *x* in the tree is one of its children who is closer to the parent of *x*.

References

1. H. Alemdar, C. Ersoy, Wireless sensor networks for healthcare: a survey. Comput. Netw. **54**(15), 2688–2710 (2010)
2. L. Atzori, A. Iera, G. Morabito, The internet of things: a survey. Comput. Netw. **54**(15), 2787–2805 (2010). doi:10.1016/j.comnet.2010.05.010
3. G. Bag, M.T. Raza, K.H. Kim, S.W. Yoo, LoWMob: intra-PAN mobility support schemes for 6LoWPAN. Sensors **9**(7), 5844–5877 (2009)
4. M. Bouaziz, A. Rachedi, A survey on mobility management protocols in wireless sensor networks based on 6LoWPAN technology. Comput. Commun. **52** (2014). doi:10.1016/j.comcom.2014.10.004
5. N. Bui, M. Zorzi, Health care applications: a solution based on the internet of things, in *Proceedings of the 4th International Symposium on Applied Sciences in Biomedical and Communication Technologies*, pp. 131:1–131:5 (2011)
6. E. Callaway, P. Gorday, L. Hester, J.A. Gutierrez, M. Naeve, B. Heile, V. Bahl, Home networking with IEEE 802.15.4: a developing standard for low-rate wireless personal area networks. Comm. Mag. **40**(8), 70–77 (2002)
7. A. Dohr, R. Modre-Opsrian, M. Drobics, D. Hayn, G. Schreier, The internet of things for ambient assisted living, in *2010 17th International Conference on Information Technology: New Generations (ITNG)*, pp. 804–809 (2010). doi:10.1109/ITNG.2010.104
8. J. Eriksson, F. Österlind, N. Finne, N. Tsiftes, A. Dunkels, T. Voigt, R. Sauter, P.J. Marrn, COOJA/MSPSim: interoperability testing for wireless sensor networks, in *Proceedings of the 2nd International Conference on Simulation Tools and Techniques*, pp. 1–7 (2009). doi:10.4108/ICST.SIMUTOOLS2009.5637
9. H. Fotouhi, M. Alves, A. Koubaa, N. Baccour, On a reliable handoff procedure for supporting mobility in wireless sensor networks, in *9th International Workshop on Real-Time Networks* (2010)
10. V. Gazis, K. Sasloglou, N. Frangiadakis, P. Kikiras, Wireless sensor networking, automation technologies and machine to machine developments on the path to the internet of things, in *2012 16th Panhellenic Conference on Informatics (PCI)*, pp. 276–282 (2012)
11. M. Ha, D. Kim, S.H. Kim, S. Hong, Inter-mario: a fast and seamless mobility protocol to support inter-pan handover in 6LoWPAN, in *2010 IEEE Global Telecommunications Conference (GLOBECOM 2010)*, pp. 1–6 (2010)
12. S. Hong, D. Kim, M. Ha, S. Bae, S.J. Park, W. Jung, J.E. Kim, SNAIL: an IP-based wireless sensor network approach to the internet of things. IEEE Wirel. Commun. **17**(6), 34–42 (2010)
13. M.M. Islam, E.N. Huh, Sensor proxy mobile IPv6 (SPMIPv6)-a novel scheme for mobility supported IP-WSNs. Sensors **11**(2), 1865–1887 (2011). doi:10.3390/s110201865
14. A. Jabir, S. Subramaniam, Z. Ahmad, N. Hamid, A cluster-based proxy mobile IPv6 for IP-WSNs. EURASIP J. Wirel. Commun. Netw. (1) (2012). doi:10.1186/1687-1499-2012-173
15. N. Khalil, M. Abid, D. Benhaddou, M. Gerndt, Wireless sensors networks for internet of things, in *2014 IEEE 9th International Conference on Intelligent Sensors, Sensor Networks and Information Processing (ISSNIP)*, pp. 1–6 (2014)
16. C. Koop, R. Mosher, L. Kun, J. Geiling, E. Grigg, S. Long, C. Macedonia, R. Merrell, R. Satava, J. Rosen, Future delivery of health care: cybercare. IEEE Eng. Med. Biol. Mag. **27**(6), 29–38 (2008). doi:10.1109/MEMB.2008.929888

17. P. Kulkarni, Requirements and design spaces of mobile medical care. ACM SIGMOBILE Mob. Comput. Commun. Rev. **11**(3), 12–30 (2007)
18. S. Kumara, L. Cui, J. Zhang, Sensors, networks and internet of things: research challenges in health care, in *Proceedings of the 8th International Workshop on Information Integration on the Web: In Conjunction with WWW 2011*, pp. 2:1–2:4 (2011)
19. L. Mainetti, L. Patrono, A. Vilei, Evolution of wireless sensor networks towards the internet of things: a survey, in *2011 19th International Conference on Software, Telecommunications and Computer Networks (SoftCOM)*, pp. 1–6 (2011)
20. D. Miorandi, S. Sicari, F.D. Pellegrini, I. Chlamtac, Internet of things: vision, applications and research challenges. Ad Hoc Netw. **10**(7), 1497–1516 (2012)
21. J. Montavont, D. Roth, T. No, Mobile IPv6 in internet of things: analysis, experimentations and optimizations. Ad Hoc Netw. **14**, 15–25 (2014). doi:10.1016/j.adhoc.2013.11.001
22. M.S. Shahamabadi, B.B.M. Ali, P. Varahram, A.J. Jara, A network mobility solution based on 6LoWPAN hospital wireless sensor network (NEMO-HWSN), in *Proceedings of the 2013 7th International Conference on Innovative Mobile and Internet Services in Ubiquitous Computing, IMIS '13*, pp. 433–438 (2013). doi:10.1109/IMIS.2013.157
23. X. Shang, R. Zhang, F. Chu, An inter-PAN mobility support scheme for IP-based wireless sensor networks and its applications. Inf. Technol. Manag. **14**(3), 183–192 (2013). doi:10.1007/s10799-013-0155-z
24. X. Wang, D. Le, H. Cheng, Mobility management for 6LoWPAN wireless sensor networks in critical environments. Int. J. Wirel. Inf. Netw. **22**(1), 41–52 (2015)
25. X. Wang, D. Le, Y. Yao, C. Xie, Location-based mobility support for 6LoWPAN wireless sensor networks. J. Netw. Comput. Appl. **49**, 68–77 (2015)
26. X. Wang, C.H. Le Deguang, C. Xie, All-IP wireless sensor networks for real-time patient monitoring. J. Biomed. Inf. **52**, 406–417 (2014). doi:10.1016/j.jbi.2014.08.002
27. L.D. Xu, W. He, S. Li, Internet of things in industries: a survey. IEEE Trans. Ind. Inf. **10**(4), 2233–2243 (2014). doi:10.1109/TII.2014.2300753
28. A. Zamanifar, E. Nazemi, M. Vahidi-Asl, DSHMP-IOT: a distributed self healing movement prediction scheme for internet of things applications. Appl. Intell. 1–21 (2016)
29. Q. Zhu, R. Wang, Q. Chen, Y. Liu, W. Qin, IOT gateway: bridgingwireless sensor networks into internet of things, in *2010 IEEE/IFIP 8th International Conference on Embedded and Ubiquitous Computing (EUC)*, pp. 347–352 (2010). doi:10.1109/EUC.2010.58
30. Z. Zinonos, V. Vassiliou, Inter-mobility support in controlled 6LoWPAN networks, in *2010 IEEE GLOBECOM Workshops (GC Wkshps)*, pp. 1718–1723 (2010). doi:10.1109/GLOCOMW.2010.5700235

The manufacturer's authorised representative in the EU is Springer Nature Customer Service Centre GmbH, Europaplatz 3, 69115 Heidelberg, Germany. If you have any concerns regarding our products, please contact ProductSafety@springernature.com

Printed and bound by CPI Group (UK) Ltd, Croydon, CR0 4YY
29/07/2026
02180520-0015